T0224968

# SpringerBriefs in Applied Sciences and Technology

SpringerBriefs present concise summaries of cutting-edge research and practical applications across a wide spectrum of fields. Featuring compact volumes of 50 to 125 pages, the series covers a range of content from professional to academic.

Typical publications can be:

- A timely report of state-of-the art methods
- An introduction to or a manual for the application of mathematical or computer techniques
- A bridge between new research results, as published in journal articles
- A snapshot of a hot or emerging topic
- An in-depth case study
- A presentation of core concepts that students must understand in order to make independent contributions

SpringerBriefs are characterized by fast, global electronic dissemination, standard publishing contracts, standardized manuscript preparation and formatting guidelines, and expedited production schedules.

On the one hand, **SpringerBriefs in Applied Sciences and Technology** are devoted to the publication of fundamentals and applications within the different classical engineering disciplines as well as in interdisciplinary fields that recently emerged between these areas. On the other hand, as the boundary separating fundamental research and applied technology is more and more dissolving, this series is particularly open to trans-disciplinary topics between fundamental science and engineering.

Indexed by EI-Compendex, SCOPUS and Springerlink.

Suzairi Daud

# Carbon Nanotubes

Fabrication Using the Arc Discharge Process

 Springer

Suzairi Daud
Department of Physics,
Faculty of Science
Universiti Teknologi Malaysia
Johor Bahru, Johor, Malaysia

ISSN 2191-530X                    ISSN 2191-5318  (electronic)
SpringerBriefs in Applied Sciences and Technology
ISBN 978-981-99-4961-8           ISBN 978-981-99-4962-5  (eBook)
https://doi.org/10.1007/978-981-99-4962-5

This Springer imprint is published by the registered company Springer Nature Singapore Pte Ltd.
The registered company address is: 152 Beach Road, #21-01/04 Gateway East, Singapore 189721,
Singapore

*Alhamdulillah... Praised be to Allah s.w.t...
Peace and Blessing be to Prophet
Muhammad s.a.w...
First and foremost, I would like to extend my
heartfelt gratitude to all the individuals who
have directly and indirectly contributed to the
preparation and completion of this book.
Countless researchers and academicians
have been involved, providing their
invaluable insights and expertise in the fields
of physics, fiber sensor technology, carbon
nanotubes, and beyond.
In particular, I would like to express my
deepest appreciation and thanks to
academicians and staff of Universiti
Teknologi Malaysia for their support and
collaboration throughout this endeavor. Their
contribution has been instrumental in
shaping my understanding and thoughts on
the subject matter.
I would also like to extend my sincere
appreciation to my beloved wife, my
wonderful children, my late father, my
mother, my siblings, my esteemed colleagues,
my students, and all my dear friends. Their
unwavering collaboration, encouragement,*

*guidance, motivation, support, patience, and friendship have been crucial throughout this journey. Without their continuous support and unwavering interest, the completion of this book would have been an insurmountable task.*

*Once again, I am profoundly grateful to each and every individual who has played a part in bringing this book to fruition. Your contributions have been invaluable, and I am truly indebted to you all.*

# Foreword

It is my distinct pleasure to introduce this groundbreaking book, *Carbon Nanotubes: Fabrication Using Arc Discharge Process*, authored by Ts. Dr. Suzairi Daud from Universiti Teknologi Malaysia.

Within the pages of this comprehensive work, the Dr. Suzairi unveils the intricacies of fabricating carbon nanotubes using arc discharge plasma, a technique that has revolutionized the field of nanomaterials. With unwavering dedication and a wealth of expertise, he guides us through the nuanced processes, experimental methodologies, and theoretical foundations that underpin the synthesis of these remarkable structures.

From the fundamentals of arc discharge plasma to the optimization of synthesis parameters, each chapter is meticulously crafted to provide readers with a comprehensive understanding of the subject matter.

This book is not merely a compilation of research findings. It is a testament to the immense potential of carbon nanotubes and the groundbreaking advancements that can be achieved through meticulous fabrication techniques.

The impact of this book extends far beyond academia. Its insights into the fabrication of carbon nanotubes using arc discharge plasma hold immense promise for a myriad of applications, from nanoelectronics and energy storage to biomedical engineering and beyond. By shedding light on the intricacies of this cutting-edge technology, the author paves the way for future innovation and the translation of scientific discoveries into real-world solutions.

It is no doubt that *Carbon Nanotubes: Fabrication Using Arc Discharge Process* book will serve as a guiding light for researchers, students, and professionals alike, inspiring them to delve deeper into the realm of carbon nanotube fabrication and explore the limitless possibilities that lie ahead.

Johor Bahru, Malaysia                                                      Suzairi Daud

# Preface

Welcome to the captivating world of carbon nanotubes and the remarkable technique of fabrication using arc discharge plasma. It is with great excitement and a sense of awe that I present this book, which explores the intricacies, advancements, and potential applications of this groundbreaking field.

As a researcher and enthusiast in the realm of nanotechnology, my journey into the realm of carbon nanotubes has been both exhilarating and humbling. These cylindrical wonders have captivated the scientific community with their extraordinary properties, holding immense promise for advancements in various fields, from electronics and energy storage to biomedicine and materials science.

In the following pages, I endeavor to unravel the intricacies of fabricating carbon nanotubes using arc discharge plasma, an innovative technique that has revolutionized the synthesis process. Drawing upon years of research, experimentation, and collaboration with brilliant minds in the field, I aim to present a comprehensive guide that combines theoretical foundations, practical methodologies, and the latest advancements in this rapidly evolving area of study.

This book is divided into five distinct chapters: Introduction, Theory and Operational Principles of Carbon Nanotubes, Research Design, Production of Carbon Nanotubes Using Arc Discharge Plasma, and Conclusion.

It begins by providing a concise explanation of carbon nanotubes and outlines the scope and limitations of the book, which is discussed in detail in Chap. 1.

Chapter 2 delves into the theory and operational principles of carbon nanotubes. It provides an in-depth review of carbon ions present in arc discharge plasma, explores the properties of carbon nanotubes and carbon ions, and discusses the characteristics of hydrogen and air. Additionally, it comprehensively covers topics such as the review of existing carbon nanotube research, CR-39 track detector analysis, the operational principles of Nd:YAG laser, and plasma generation.

Chapter 3 focuses on the research design of the study. It provides detailed information about the experimental equipment used and describes the experimental setup required to carry out the research effectively.

In Chap. 4, the production of carbon nanotubes using arc discharge plasma is thoroughly explained. The chapter begins with an examination of the etching process and proceeds to present the findings of the arc discharge process for carbon nanotube production under different pressures and surroundings.

Finally, in Chap. 5, the book presents the conclusion of the research and offers recommendations for future work. This chapter serves as a comprehensive summary of the study's key findings and implications, providing insights into the potential avenues for further exploration in the field of carbon nanotube fabrication using arc discharge plasma.

By organizing the book in this manner, readers are guided through a logical progression from introduction to conclusion, ensuring a comprehensive understanding of the subject matter. It is my hope that this book serves as a valuable resource for researchers, scientists, and students alike, inspiring further advancements and innovations in the exciting field of carbon nanotube fabrication.

Johor Bahru, Malaysia                                                         Suzairi Daud

# Acknowledgements

I would like to express my sincere gratitude and appreciation to all those who have supported and contributed to the completion of this book.

First and foremost, I would like to extend my deepest thanks to all academicians, staff, and colleague from Faculty of Science, Universiti Teknologi Malaysia, for the invaluable guidance, expertise, and encouragement throughout this journey. I am very grateful for their stimulating discussions, valuable insights, and collaborative spirit

To my family and friends, I am profoundly grateful for their unwavering support, understanding, and encouragement throughout this endeavor. Their love and belief in me have been a constant source of motivation, and I am truly fortunate to have them by my side.

Finally, I want to acknowledge the countless researchers, scholars, and authors whose work has paved the way for this. Their contributions have been instrumental in shaping my understanding and providing a solid foundation for this research.

While it is not possible to name everyone individually, please know that your contributions, whether big or small, have been deeply valued and appreciated. This book would not have been possible without the collective support, guidance, and collaboration of all those mentioned and the many others who have contributed in various ways.

Thank you all for being a part of this book.

# Contents

# Abbreviations

| | |
|---|---|
| ac | Alternating current |
| ADC | Allyl diglycol carbonate |
| BPO | Benzoyl peroxide |
| CBs | Carbon blacks |
| CIRT | Carbon ion radiotherapy |
| CNFs | Carbon nanofibers |
| CNTs | Carbon nanotubes |
| CNWs | Carbon nanowalls |
| CRTs | Cathode ray tubes |
| dc | Direct current |
| DLC | Diamond-like carbon |
| DNA | Deoxyribonucleic acid |
| ESD | Electrostatic discharge |
| i.e. | In example |
| IC | Integrated circuit |
| LET | Linear energy transfer |
| LIBS | Laser-induced breakdown spectroscopy |
| MWCNTs | Multi-walled carbon nanotubes |
| NaOH | Sodium hydroxide |
| Nd:YAG | Neodymium-doped yttrium aluminum garnet |
| NIRS | National Institute of Radiologic Sciences |
| RBE | Relative biological effectiveness |
| ROS | Radical oxygen species |
| SSNTD | Solid-state nuclear track detector |
| SSNTDs | Solid-state nuclear track detectors |
| STP | Standard temperature and pressure |
| SWCNTs | Single-walled carbon nanotubes |
| YLF | Yttrium lithium fluoride |
| YVO4 | Yttrium orthovanadate |

# Chapter 1
# Introduction

## 1.1 Overview

In Physics, a state of matter is one of the distinct forms in which matter can exist. Four states of matter are observable in everyday life, which are solid, liquid, gas, and plasma. Many intermediate states were known to exist, such as liquid crystal, and some states only exist under extreme conditions, such as Bose–Einstein condensates, neutron-degenerate matter, and quark-gluon plasma, which only occurred in situations of extreme cold, extreme density, and extremely high energy, respectively.

A solid holds a definite shape and volume without container. The particles are held very close to each other and have very strong bonding. Solid can be divided into four, which are amorphous solid, crystalline solid, plastic crystal, and quasi-crystal. Amorphous solid is a solid in which there is no far-range order of the positions of the atoms. Crystalline solid is a solid in which atoms, molecules, or ions are packed in regular order. Plastic crystal is a molecular solid with long-range positional order, but with constituent molecules retaining rotational freedom. And quasi-crystal is a solid in which the positions of the atoms have long-range order, but not in a repeating pattern.

Liquid is a mostly non-compressible fluid. Liquid is able to conform to the shape of its container but retains a nearly constant volume and independent of pressure. As the only state with a definite volume but no fixed shape, liquid is made up of tiny vibrating particles of matter (such as atoms) and held together by intermolecular bonds. A liquid is able to flow and take the shape of a container.

Most liquids resist compression, although others can be compressed. A distinctive property of the liquid state is surface tension, leading to wetting phenomena. Water is the most common liquid on Earth. Liquid can be classified into two, which are liquid crystal and non-Newtonian fluid. Liquid crystal properties are intermediate between liquids and crystals. Generally, it is able to flow like a liquid but exhibiting long-range order. While non-Newtonian fluid is a fluid that does not follow Newton's law of viscosity.

S. Daud, *Carbon Nanotubes*, SpringerBriefs in Applied Sciences and Technology, https://doi.org/10.1007/978-981-99-4962-5_1

Gas is a compressible fluid. Gas will not only conform to the shape of its container, but it will also and able to expand to fill the container. A pure gas may be made up of individual atoms (i.e., a noble gas like neon), elemental molecules made from one type of atom (i.e., oxygen), or compound molecules made from a variety of atoms (i.e., carbon dioxide). A gas mixture (such as air) contains a variety of pure gases.

What distinguishes a gas from liquids and solids is the vast separation of the individual gas particles itself. This separation usually makes a colorless gas invisible to the human observer. The gaseous state of matter occurs between the liquid and plasma states, the latter of which provides the upper temperature boundary for gases. Bounding the lower end of the temperature scale lies degenerative quantum gases which are gaining increasing attention. High-density atomic gases super-cooled to very low temperatures are classified by their statistical behavior as either Bose gases or Fermi gases.

The only chemical elements that are stable diatomic homonuclear molecules at standard temperature and pressure (STP) are hydrogen ($H_2$), nitrogen ($N_2$), oxygen ($O_2$), and two halogens: fluorine ($F_2$) and chlorine ($Cl_2$). When grouped together with the monatomic noble gases like helium (He), neon (Ne), argon (Ar), krypton (Kr), xenon (Xe), and radon (Rn), these gases are referred as elemental gases.

Meanwhile, plasma is free charged particles and usually in equal numbers, such as ions and electrons. Unlike gases, plasma may self-generate magnetic fields and electric currents and respond strongly and collectively to electromagnetism forces. Plasma is very uncommon on Earth (except for the ionosphere), although it is the most common state of matter in the universe.

As discussed earlier, plasma is one of the four fundamental states of matter and was first described by chemist Irving Langmuir in the 1920s. It consists of a gas of ions (atoms which have some of their orbital electrons remove) and free electrons. Plasma is the fourth state of matter which is a conglomerate of neutron, electrons, and ions. The study of laser-induced carbon plasma and other carbon plasmas by spectroscopic means becomes important in materials science field especially for laser deposition of various carbon forms and carbon nanostructure.

The interaction between high-energy laser pulse and material leads to the formation of plasma through laser ablation which is used for variety of technological applications. The energy from laser pulse is absorbed by target material, and it increases the surface temperature and leads to ablation, evaporation, and ionization of the target material. The laser-induced carbon plasma is used to grow a variety of carbon-related materials such as fullerenes, diamond-like carbon thin films, and carbon nanotubes.

Carbon ions design a graphite-like structure which implies to develop the nano-hardness and tribological properties. The study of laser-induced carbon plasma is important indeed. Optimizing the parameters is the best way to observe the properties of carbon ions.

The determination of the carbon ion's energy under different ambient environments and pressures by using CR-39 target is very much important. The parameters used such as laser source, target materials, substrate of CR-39, and plasma dynamics have direct impact on the properties on the determination of carbon ions itself. Thus, it is important to study the carbon ions behavior to optimize the properties of carbon ions for different technological applications.

Ablation of material from the target surface starts soon after the laser pulse interacts with the target surface, which leads to the formation of plasma (also called laser-induced plasma). Nd:YAG laser is among the best laser source used to create carbon plasma by focusing it on the graphite surface materials. The plasma with high temperature and electron density was formed due to the vaporization and excitation of material.

In general, CR-39 material is exposed to energetic carbon ions under different ambient conditions in order to investigate the properties of carbon ions by analyzing the ions track on CR-39 material. CR-39 detector, or also known as solid-state nuclear track detectors (SSNTDs), is a class of plastic detector. It is widely used for the registration of heavy charged particles.

In addition, the CR-39 detector is suitable for protons, deuterons, and alpha detection because of their high sensitivity toward the particle. Due to their structural stability, optical clarity, and heat resistance, it is very suitable for industrial, medical, and optical uses. SSNTDs also give beneficial results in numerous fields such as geophysics, nuclear physics, and astrophysics. The characterization of carbon ions in arc discharge plasma is important indeed. Solid-state nuclear track detector is one of the mediums that can be used for the characterization of carbon ions in arc discharge plasma.

Meanwhile, carbon nanostructure is very unique carbon elements. It can form a ball-shaped fullerenes and the cylindrical nanotubes. Buckyballs, polyaromatic molecules, and graphene are made of C60 material. The carbon nanostructures include various low-dimension allotropes of carbon including carbon nanotubes and others.

Carbon nanostructures become more popular because of the fullerenes and nanotubes. There are also nanocarbon families that increase the current attention. Nanoscale region is the characteristic sizes of carbon structures. Diamond-based materials at the nanoscale range from single diamond clusters to bulk nanocrystalline films, and this is the variety use of nanodiamond. Carbon blacks (CBs), fullerenes, carbon nanotubes, and carbon fibers are the examples of carbon nanomaterials that are related to nanotechnology application. Energy and materials segments can be covered by applications of carbon nanomaterials.

Carbon nanostructures can be produced using many methods such as laser ablation method, solar technique, and arc discharge process. In this book, arc discharge process was chosen as the medium to produce the carbon nanostructures, and it will be discussed in detail from the theory of the formation of carbon nanotubes till the experimental process of the production of carbon nanotubes based on the arc discharge process that is discussed.

Arc discharge is generated between two electrodes under inert atmosphere. Sublimation of carbon will happen between the two rods due the high temperature. The discharge gap of an arc discharge is filled with a plasma that consists of electrons, ions, and few neutral elements. It also consists of excited atoms and molecules of the working gas, including the electrode materials. For the temperature of an arc discharge, a distinction is made among the ion temperature, the electron temperature, and the temperature of the neutral component.

Characteristics of carbon ions are determined using the solid-state nuclear track detector (SSNTD) made by CR-39 materials. SSNTD is dielectric materials, crystalline or vitreous, which can track of charged nuclear particles, like alpha particles or fission fragments. The etched nuclear particles track is faster than the bulk material, and the size and shape of these can give the information about the mass, charge, energy, and direction of motion of the particles.

The detectors origin tracks the chemical etching that is visible at the optical microscope. The track etching rate is higher along the latent track, where damage due to the charged particle increases the chemical potential, and etching rate gives rise to holes, which is the etched tracks. CR-39 has very high sensitivity of detection toward the particles. The information on individual particles from the radiation detectors, the persistence of the tracks allowing measurements to be made over long periods of time, and the simple, cheap, and robust construction of the detector information can be determined.

The main motivation of this book is to discuss on the formation and energy of carbon ions in plasma induced by Nd:YAG laser under different ambient conditions, including environment and pressure. The determination of the energy of carbon ions using CR-39 track detector in different ambient environment (air and hydrogen) and ambient pressure (0.1, 1, 10, and 100 mbar) will be discussed in detail.

This book focuses on the detection and determination of the carbon ions and their energy using CR-39 track detector and graphite as the laser ablation target. A Q-switched Nd:YAG laser with 1064 nm wavelength, 740 mJ energy, 10 Hz frequency, and 6 ns pulse duration is used to generate carbon laser-induced plasma for the purposes.

## 1.2  Scopes of the Book

Arc discharge is a crucial method for fabricating carbon nanostructures, utilizing graphite as one of the primary carbon sources for generating arc discharge plasma. This process creates a graphite-like structure, enhancing the nano-hardness and tribological properties of the carbon ions. Despite significant progress, research on the deposition of carbon forms and nanostructures is still ongoing.

Numerous methods have been employed to produce carbon ions, including their use in cancer therapy. Optimizing parameters such as laser source, target material, CR-39 substrates, and arc discharge plasma can directly impact the production of carbon ions. Therefore, studying carbon ions under different ambient environments and pressures is essential to obtain valuable information on their state and energy.

To determine the state and energy of carbon ions, they will be generated under different ambient environments, such as air and hydrogen, and at different pressures, which are 0.1, 1, 10, and 100 mbar, respectively.

The SSNTD material will then be etched in 6.25 M of NaOH solution for seven hours at $(70 \pm 1)$ °C, and the nuclear track will be observed under an AmScope optical microscope. The study will focus on detecting the carbon ions' energy and their reaction with CR-39 using solid-state nuclear track detectors (SSNTDs).

# Chapter 2
# Theory and Operational Principles of Carbon Nanotubes

## 2.1 Introduction

This chapter provides an in-depth background and literature review on carbon nanotubes (CNTs), including their applications. The theoretical aspects and operational principles of CNTs will be thoroughly discussed. It commences with a brief introduction to CNTs and emphasizes the critical aspects of carbon ions, laser-induced carbon plasma, CR-39 track detectors, and other related topics. Additionally, this chapter will delve into the working principles of Nd:YAG lasers and the various applications of laser-induced carbon plasma. All of these topics will be explained in detail, providing a comprehensive understanding of CNTs.

## 2.2 Review of Carbon Ion in Arc Discharge Plasma

The growth of carbon nanotubes has been studied through some aspect to gain the availability of the CNT growth techniques. Graphite rods which are in contact with applying alternated current (AC) voltage in an inert gas were evaporated. Direct current (DC) arc voltage was applied between two separated graphite rods by modifying the SiC powder production apparatus and fullerences were generated by anode fullerenes in the form of soot in the chamber, and a part of the evaporated anode is deposited on the cathode. In that cathode deposit, CNTs were found.

Arc DC voltage was applied between two graphite rods; after evacuating the chamber with a vacuum pump, an appropriate ambient gas is introduced at the desired pressure. When the pure graphite rods were used, the anode evaporates to form fullerenes, which are deposited in the form of soot in the chamber.

CNTs were included when a small part of the evaporated anode was deposited on the cathode. In order to clarify the effect of gas (including hydrogen atom) in multi-walled nanotubes production, ambient $CH_4$ gas was analyzed before and after arc discharge process by mass spectroscopy.

S. Daud, *Carbon Nanotubes*, SpringerBriefs in Applied Sciences and Technology,
https://doi.org/10.1007/978-981-99-4962-5_2

Hydrogen gas was used to evaporate pure graphite rods. A few coexisting carbon nanoparticles can be seen through scanning electron micrograph after MWNTs produced by hydrogen arc discharge (H2-arc MWNTs). These nanoparticles are easily removed by infrared irradiation or heating in air at 500 °C temperature. Inner diameter of these nanoparticles is typically as low as 7 Å, which is equal to the diameter of C60 was obtained. The 3 Å tubes are the latest MWNTs found to exist inside H2-arc.

SWNTs were also the first produced by arc discharge using a graphite anode containing a metal catalyst (Fe or Co). However, SWNTs were not obtained from the cathode deposit but were obtained from the soot in the gas phase. Mass production of SWNTs by arc discharge was achieved by using a bimetallic Ni-Y catalyst in He ambient gas. The method was effectively modified by using two graphite electrodes inclined at an angle of 30 ° instead of the conventional 180 ° alignment.

Fe catalyst instead of Ni-Y and an H2-Ar gas mixture in place of He also can be used for another method. It produces a partly aligned macroscopic net of SWNTs as long as 30 cm. By heating in air at 420 °C, SWNTs are easily purified and then rinsed in mild HCl.

CNTs were found hard to grow by arc discharge. Arc discharge method has higher growth temperature than other method of CNTs production. CNTs produced by arc discharge are generally high of crystallinity and perfection, and the yield per unit time is also higher than other methods.

Fullerenes, nanotubes, and graphene are structurally similar. They all consist of a network of SP2 carbon atoms that typically results from the condensation of elementary carbon atoms in the gas phase or at the surface of a catalyst. The catalyzed synthesis of nanocarbon is the most popular synthesis technique, as it enhances the overall conversion of the carbon precursor and offers the possibility to improve the selectivity of the reaction by tuning the catalyst as well as the reaction conditions.

In addition, it benefits from existing knowledge, gained from surface science, on the interaction of hydrocarbons with transition metal surfaces, and from the pioneering work of Baker, Oberlin, and Endo on the growth of carbon fibers.

The synthesis of CNTs (both single and multi-walled) largely benefited from the knowledge gained on important industrial petrochemical processes. Many of these are catalyzed by transition metals such as Fe, Ni, or Co. The interaction of these metals, either as single crystals or as nanoparticles, with small hydrocarbons (e.g., $CH_4$, $C_2H_2$, $C_2H_4$) has been extensively studied in order to improve the reactivity and hence performance of the catalysts.

A study has been made to measure the bulk etch rates of the CR-39 for different molarities of NaOH/ethanol through the masking method and to measure the depths of pre-etched tracks which are further etched in NaOH/ethanol. The effect of stirring on the bulk etch rate also being studied.

If the production rate of sodium carbonate increases to a point such that the sodium carbonate can effectively insulate the detector from the etchant, and the bulk etch rate will decrease if the molarity of NaOH/ethanol keeps increasing.

The height difference between the etched portion and the masked portion that masked by epoxy can be used to determine the resistance to etching through surface profilometry measurements for the bulk etch rates of the CR-39. The effect of stirring on the bulk etch rate of the CR-39 detector also a concern, so the bulk etch rates were determined under the conditions with and without stirring during etching, and magnetic stirrer was provided to stir.

For the depths of pre-etched, tracks will produce two type of periods pre-etched, which are short and long period. The tracks will also shorten when etched by NaOH. There are three steps that need to be followed.

Firstly, irradiation of CR-39 detectors. The original dimensions of the sheet of detector are $(30 \times 47 \times 0.1)$ cm thickness. The detectors were cut into the size of $(1 \times 1)$ cm$^2$ and were irradiated with energies of 1, 2, 3, 4, and 5 MeV, respectively. The alpha source was employed, and the final alpha energies incident on the detector and detector distances in normal air were used to control the source. The relationship between the alpha energy and the air distance traveled by an alpha particle with initial energy of 5.49 MeV from 241 Å was obtained by measuring the energies for alpha particles passing different distances through normal air using alpha spectroscopy systems.

Secondly, pre-etching in NaOH/water and further etching in NaOH/ethanol. To reveal one set of the detectors, it were pre-ached in 6.25 M of NaOH/water solutions and maintained at 70 °C in a water bath for 1.5, 2, and 4 h (short pre-etching periods) for incident alpha energies of 1, 2, 3, 4, and 5 MeV, respectively. After the pre-etching process, the NaOH solutions was dissolved in absolute ethanol (95 %) to prepare the etchants. The etching temperature was chosen to be at 55 °C.

Thirdly, measurement of track depths. The replica method was used to measure the track lengths for spherical tracks. The replica method is preferred as surface profilometry can provide a better spatial resolution (up to 40 nm) than that provided by general optical microscopes (0.3 μm). The track depths were determined from the height difference between the top and the base of tracks in the lateral view of the image of replica.

New empirical equation describing the charged particles radiation track development against etching time and track longitudinal depth is presented. Parameters' values obtained from experimental data can be used to predict etched track lengths at different energies and etching times. The sensitivity of CR-39 is strongly influenced by the treatment of detector samples before, which are during and after the exposure and the final evaluation process by chemical etching.

Whereas changes in detection properties by external environmental influences are generally considered, the dependences on the etching conditions are ignored. Commonly, the sensitivity is assumed to compensate variations in the etching conditions for track revealing. In the frame of the existing database, the sensitivity is not really independent on variations in etching temperatures and should be corrected for differences in the activation energies for stimulation of the bulk and track etching process.

An electrical discharge results from the creation of a conducting path between two points of different electrical potential in the medium in which the points are immersed. For a discharge to occur, there must usually be a source of electrons at the cathode, and the nature of this source controls the form of the discharge.

The arc discharge is a high current and low voltage discharge, in contrast with the low current, high voltage glow discharge. It is characterized by a negative resistance VI characteristic and high temperatures. Electrons for the discharge were supplied by a cathode spot that much more efficient electron emitter than the glow discharge cathode phenomena. High-field emission is essentially quantum mechanical tunneling through the potential barrier at the surface of the cathode.

Synthesis of carbon nanowalls (CNWs) at atmospheric pressure is realized by using a negative normal glow discharge, which differs from prior low-pressure plasma-enhanced chemical vapor deposition methods for CNWs growth and holds great potential for mass production of CNWs. The CNWs growth process is examined for the effect of discharge regime, discharge duration (growth time), hydroxyl radicals (water vapor), and local current density.

OH radicals play an essential role in the initial nucleation process, but excess OH radicals accelerate the oxidation of CNWs. For a fixed growth time, there exists an optimum feed gas relative humidity and an optimum current density of $\sim 40 \%$ and $\sim 9.17$ A/m$^2$, respectively, for the growth of large and less defective CNWs with a high degree of crystallinity.

Low-temperature non-equilibrium plasma was generated during gliding arc discharge in a multi-electrode reactor at atmospheric pressure. In the case of gas temperature, which is much lower in temperature, it was determined directly by the thermo vision camera to avoid tedious calculations from the spectral lines.

The results of the plasma diagnostics generated by gliding arc for different process gases and their flow rates are presented and discussed. Simplified diagnostic method would be very perspective for monitoring various technological processes where plasma is used such as treatment of flue gases, wastes utilization, deodorization, disinfection and sterilization, material processing, and new material manufacturing for application in microelectronics and nanotechnologies.

The arc can operate either in the ambient gas or in the vapor emitted by the cathode and anode as they vaporize. The electrodes operate at their boiling points, which determine the maximum temperatures available. The voltage of a carbon arc in air is given approximately by $V = 38.8 + 2.0 + (16.6 + 10.5x)/I$ where $V$ is the voltage in volts (V), $I$ is the current flow in ampere (A), and $x$ is the arc length measured in millimeters (mm).

The cathode is heated by positive ion bombardment, where the effect of the ions is purely thermal and any positive ion induced electron emission is unimportant. There must always be sufficient ions to keep the cathode hot.

The field at the surface of the cathode can be estimated as $E = \frac{4V}{3d}$, where $V$ is the cathode drop and $d$ is the width of the cathode fall region. The positive column differs significantly between arcs operated at low pressures (say, below 10 cm Hg pressure) and arcs at higher pressures, such as atmospheric.

At low pressures, it is like the positive column of a glow discharge, with a very high electron temperature (40,000 K is not unusual) and a low ion and gas temperature (say, 300 K). At high pressures, electron, ion, and gas temperatures are equal and high, and ionization is principally thermal.

It is shown that characteristics of synthesized SWNTs can be altered by varying plasma parameters. Effects of electrical and magnetic fields applied during SWNT synthesis in arc plasma are explored. Magnetic field has a profound effect on the diameter, chirality, and length of a SWNT synthesized in the arc plasma.

An average length of SWNT increases by a factor of 2 in discharge with magnetic field, and a number of long nanotubes with the length above 5 μm also increases in comparison with that observed in the discharge without a magnetic field. In addition, synthesis of a few-layer graphene in a magnetic field presence is discovered.

Calculations indicate that substantial fraction of the current at the cathode is conducted by ions (0.7 – 0.9 of the total current). It is shown that non-monotonic behavior of the arc current-voltage characteristic can be reproduced taking into account the experimentally observed dependence of the arc radius on arc current.

Welding is an important application for arcs. Although oxyacetylene flames are hot enough to make fusion welds and are very useful, arc welding is relied upon for heavy duty welding. Inert gas welding uses a tungsten electrode bathed in an inert gas such as helium to avoid oxidizing the work and a filler rod.

The atomic hydrogen then recombines on the surface of the work, creating the necessary high temperature. These methods are useful in special cases, but the most general type of arc welding is the metallic arc. The arc is struck between a welding rod and the work, with the welding rod providing the filler metal.

The study of etch pit geometry and determination of the range of particles in the detector can reveal the identity of the particles forming the tracks and also their energies. The present experiment utilized the 20 ° beam line of the REX-ISOLDE facility at CERN.

The 2.82 MeV/u, 129Xe, and 78Kr ion beams were produced in the ISOLDE GPS target and accelerated with the REX-ISOLDE linear accelerator. The 78Kr beam contained an admixture of 2.82 MeV/u 49Ti ions which were detected with PET and which provided an additional data point for calibration.

To cover all four apertures and fixed onto the downstream side of the target ladder with screws, the PET films were cut into rectangular strips (8.0 cm × 3.5 cm). A 6.25 M NaOH solution at 55.0 ± 0.5 °C was etched which was found to be the ideal combination for etching PET detectors. Leica digital microscope was used to study the sample under × 100 dry objective which was interfaced with a computer preloaded with an image analysis software.

The large etching products cannot be removed from the track channel fast enough because of the limited by the transportation of etching solution. PET can be used to get effective charged particle detector with high detection threshold. For the significant, PET has the low cost compared to SSNTDs.

Flame reactors are used to produce fullerenes currently in industry. Fullerenes display a wide range of different biological activities. Strong antioxidant capacities and effective quenching radical oxygen species (ROS) made fullerenes suitable active compounds in the formulation of skin care products. Fullerene C60 and its derivates have been subject of intensive research.

The continuous production of carbon nanostructures by employing the flame heat is the gas-phase synthesis method to initiate chemical reactions producing condensable monomers. Two graphites are used to produce fullerenes and carbon nanotubes by using arc discharge. The origin of the carbon precursor comes from the ablation of a graphite electrode.

Hydrocarbon (benzene, methane, acetylene, ethylene) has no influence on the products' properties. In that case, benzene was used as the origin of the carbon precursor, and the experimental device can be simplified. The electrical source with 50 Hz frequency and 8 kW maximum input power was connected. Argon (99.99 % purity) and benzene were used as plasma gas and as carbon source, respectively.

At the same time, plasma gas and benzene were injected at atmospheric pressure with flow rates of 80 l/h. About 380 V voltage was used to carry out the discharge with current of 6, 8, 14, and 20 A. Electron microscopy (S-4800, FE-SEM) was used to scan the characteristics of carbon nanostructures, a high-resolution transmission electron microscopy (JEM-2100, HRTEM-200 kV), powder X-ray diffraction (DX-2700, XRD), and laser Raman spectra.

These CNTs consistent with the graphite consist of more than ten concentric carbon shells with a spacing of 0.34 nm. The carbon product deposited on the electrode ends is comprised of MWCNTs, aggregate carbon nanoparticles, and agglomerate carbon particles. An arc discharge current of 8 A was obtained from the maximum yield of CNT.

## 2.3   Carbon

Carbon is a chemical element with the symbol $C$ and atomic number 6. It is nonmetallic and tetravalent, making four electrons available to form covalent chemical bonds. It belongs to group 14 of the periodic table.

Carbon is one of the softest (graphite) and hardest (diamond) materials found in nature. The 4N pure graphite is used as a target for the generation of laser-induced carbon plasma. It is an allotrope of carbon and a good conductor of electricity and the most stable form of carbon. The 4N graphite has high purity, up to 99.99 % purity, 2.267 $g/cm^3$ density, 3652–3697 °C melting points, 4200 °C boiling point, and 119–165 W/m/K thermal conductivity.

Carbon and its components are widely distributed in nature. The estimation is that carbon forms 0.032 % of the Earth's crust. Free carbon is found in big reservoirs like hard coal, an amorphous form of the element with other complex compounds of carbon-hydrogen-nitrogen.

The Earth's atmosphere contains an ever-increasing concentration of $CO_2$ and CO, form fossil fuel burning and methane ($CH_4$). No element is more essential to life than carbon. Only carbon forms strong single bonds to itself that are stable enough to resist chemical attack under ambient conditions.

This makes carbon the ability to form long chains and rings of atoms, which are the structural basis for many compounds that comprise the living cell, of which the most important is DNA. Big quantities of carbon are found in the form of compounds. Carbon is present in the atmosphere as $CO_2$ in 0.03 % in volume. Several minerals like limestone, dolomite, gypsum, and marble contain carbonates.

All the plants and live animals are formed by complex organic compounds where carbon is combined with hydrogen, oxygen, nitrogen, and other elements. The remains of live plants and animals form deposits of petroleum, asphalt, and bitumen. The natural gas deposits contain compounds formed by carbon and hydrogen.

Carbon is unique in its chemical properties because it forms a number of compo-nents superior than the total addition of all the other elements in combination with each other. The biggest group of all these components is the one formed by carbon and hydrogen.

Although the classification is not strict, carbon forms another series of compounds considered as inorganic, in a much lower number than that of the organic compounds. Elemental carbon exists in two well-defined allotropic crystalline forms, which are graphite and diamond. Other forms with little crystallinity are vegetal carbon and black fume. Chemically pure carbon can be prepared by thermal decomposition of sugar (sucrose) in absence of air.

The physical and chemical properties of carbon depend on the crystalline structure of the element. Its density fluctuates from 2.25 g/cm$^3$ for graphite to 3.51 g/cm$^3$ for diamond. The melting point of graphite is 3500 °C, and the extrapolated boiling point is 4830 °C. Elemental carbon is an inert substance, insoluble in water, diluted acids and bases, as well as organic solvents.

At high temperatures, it binds with oxygen to form carbon monoxide or dioxide. With hot oxidizing agents like nitric acid and potassium nitrate, metallic acid $C_6(CO_2H)_6$ is obtained. Among the halogens, only fluorine reacts with elemental carbon. A high number of metals combine with the element at high temperatures to form carbides.

It forms three gaseous components with the oxygen, which are carbon monoxide, CO, carbon dioxide, $CO_2$, and carbon suboxide, $C_3O_2$. CO and $CO_2$ are the most important from the industrial point of view. Carbon forms compounds with the halogens with $CX_4$ as general formula, where X is fluorine, chlorine, bromine, or iodine. At ambient temperature, carbon tetrafluoride is gas, tetrachloride is liquid, and the other two compounds are solids. The most important of all may be the dichlorodifluoromethane, $CCl_2F_2$, called Freon.

## 2.4   Carbon Nanotubes

Carbon nanotubes (CNTs) are cylindrical molecules that consist of rolled-up sheets of single-layer carbon atoms called graphene. They can be single-walled carbon nanotubes (SWCNTs) with less than 1 nm diameter or multi-walled carbon nanotubes (MWCNTs), consisting of several concentrically interlinked nanotubes, with diam-eters reaching more than 100 nm. The length of both SWCNT and MWCNT can reach to several micrometers (µm) or even millimeters (mm).

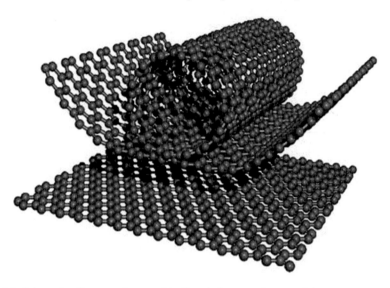

**Fig. 2.1** Schematic of how graphene could roll-up to form a carbon nanotube

Like their building block graphene, CNTs are chemically bonded with $sp^2$ bonds, an extremely strong form of molecular interaction. This feature combines with carbon nanotubes natural inclination to rope together via Van der Waals forces and provides the opportunity to develop an ultra-high strength, with low-weight materials that possess highly conductive electrical and thermal properties. This makes them highly attractive for numerous applications. Figure 2.1 shows the schematic of how graphene could roll-up to form a carbon nanotube.

The rolling-up direction of the graphene layers determines the electrical properties of the nanotubes. Chirality describes the angle of the nanotube's hexagonal carbonatom lattice. Armchair nanotubes have an identical chiral index and are highly desired for their perfect conductivity. It is so-called armchair nanotubes because of the armchair-like shape of their edges. They are unlike zigzag nanotubes, which may be semiconductors. Turning a graphene sheet, a mere 30 ° will change the nanotube forms from armchair to zigzag, or vice versa.

In the meantime, MWCNTs are always conducting and achieve at least the same level of conductivity as metals. And SWCNTs' conductivity depends on their chiral vector. They can behave like a metal and be electrically conducting, display the properties of a semiconductor, or be non-conducting. For example, a slight change in the pitch of the helicity can transform the tube from a metal into a large-gap semiconductor.

Apart from their electrical properties inherited from graphene, CNTs also have unique thermal and mechanical properties that make them intriguing for the development of new materials. Their mechanical tensile strength can be 400 times stronger than of steel. They are very lightweight, with one-sixth density of that of steel. CNTs thermal conductivity is better than that of diamond. They have a very high aspect ratio, greater than 1000. For example, in relation to their length which is extremely thin.

Their tip-surface area is near to the theoretical limit. The smaller the tip-surface area, the more concentrated the electric field, and the greater the field enhancement factor. Just like graphite, CNTs are highly chemically stable and resist virtually any chemical impact, unless they are simultaneously exposed to high temperatures and oxygen, which can make them extremely resistant to corrosion. Their hollow interior can be filled with various nanomaterials, separating, and shielding them from the surrounding environment, a property that is extremely useful for nanomedicine applications like drug delivery.

All these properties make the CNTs an ideal candidate for electronic devices, chemical/electrochemical and biosensors, transistors, electron field emitters, lithium-ion batteries, white light sources, hydrogen storage cells, cathode ray tubes (CRTs), electrostatic discharge (ESD), and electrical shielding applications.

Noted that CNTs are different from carbon nanofibers (CNFs). CNFs are usually several micrometers long and have a diameter of about 200 nm. Carbon fibers have been used for decades to strengthen compound, but they do not have the same lattice structure as CNTs.

Instead, they consist of a combination of several forms of carbon and/or several layers of graphite, which are stacked at various angles on amorphous carbon, where atoms do not arrange themselves in ordered structures. CNFs have similar properties as CNTs, but their tensile strength is lower owing to their variable structure, and they are not hollow inside.

## 2.5  Carbon Ions

Carbon ions are widely used in radiotherapy for the treatment of cancer due to their excellent in physical and biological characteristics as compared to several other types of ion's species. Physical effectiveness of the ions depends on the amount of dose, while biological effectiveness of the ions depends on the microdistribution of dose applied. In addition, heavy ions (i.e., carbon) have been used for the treatment of cancer because of their ability to provide acceptable radiation dose to the tumor.

Positively charged carbon clusters or carbon ions from laser-induced plasma was determined by Becker, and the mass spectra of both positive and negative clusters ions were clarified by calculating the binding energies of the carbon clusters itself. The formation of cluster ions in the gas phase or plasma can be observed using the mass spectroscopy method. Working with pulse length of 100 ns and repetition frequency of 5 kHz, the graphite samples were evaporated and ionized by Nd:YAG laser.

This study revealed that the impact of laser power density clearly contributes toward the distribution and formation of carbon clusters. Interestingly, LCAO-X method allowed interpretation of the abundance distribution of both positively and negatively charged cluster ions of graphite in laser-induced plasma.

It is acknowledged that carbon is an element that can crystallize in the form of diamond and graphite. There are many applications of diamond-like carbon such as magnetic hard disk coatings, wear-protective, and antireflective coatings for tribological tools, engines parts, razor blades, sunglasses, biomedical coatings, and microelectrochemical systems.

Diamond-like carbon can also be used in magnetic storage which acts as a storage device that can store the data in a thin magnetic layer called recording medium. Clusters of carbon ions have been produced by laser vaporization. Graphite has been used as a target irradiated by infrared picosecond high-intensity laser. Irradiation was performed by 35 ps wild pulse and 1064 nm wavelength of Nd:YAG laser to obtain carbon clusters ions with very high charge states and kinetic energy.

Particle therapy with protons and heavier ions like 12C has sparked renewed interest in radiation therapy in recent years. Employing heavier ions like carbon, on the other hand, has some advantages over using protons. Carbon ions have a distinct energy distribution in depth, with low levels of energy deposited in tissues close to the target and the majority of energy released in the target near the beam's end. This characteristic is known as the "Bragg curve," and the maximum energy deposition is known as the "Bragg peak."

Carbon ion radiotherapy uses a three-dimensional approach to tailor the position and shape of the Bragg peak to the tumor site. As a result, the dose can be focused on the target, while the dose to surrounding normal tissues is suppressed. As a result, a carbon ion's energy loss is 36 times larger than that of a proton traveling at the same speed, implying that localized biological damage is significantly greater, with higher cell death and fewer chances for healing.

Secondly, the ionization rate increases at the end of the particle range resulting in clusters of lesions on the DNA molecule in the cells. Given that DNA lesion clusters are much more difficult to repair than one DNA damage, carbon ions have increased biological efficiency.

In contrast to the stated advantages, the ions will partially fragmented, once they penetrate matter with energies in the MeV range. A distinctive dose tail beyond the peak will be produced when the fragments have a longer range and a wider distribution of energy than the primary ions. If the residual nuclides of carbon ions in biological materials such as tissue and bone are known, this effect can be estimated.

This carbon ion fragmentation, on the other hand, has yet to be well examined. To gain a better understanding of the generation of nuclear fragments in biological matter, further measurements and experimental information are required. When charged particles, such as protons and heavy ions, move through material, they lose energy mostly due to ionization. That energy loss is described by the Bethe-Bloch equation. A graph showing the energy loss rate, or linear energy transfer (LET), as a function of distance through a stopping medium is known as the Bragg curve. The square of the nuclear charge, $Z$, and the inverse square of the projectile velocity are the primary determinants of energy loss.

The Bragg curve takes on its famous shape as a result of this, culminating at very low energies right before the projectile comes to a halt. This Bragg peak is what distinguishes ion therapy from X-ray treatment for cancer. Below is a Bragg curve for 290 MeV/n carbon. It is possible to break up the nucleus of heavy ions when they travel through the degrader. Because the nuclear fragments all have a lower $Z$ than the main, they will have a longer range in general.

The Geant4 MC (v9.3.2) was used in a recent work to evaluate the Bragg peak for a supplied dose of carbon-12 (12C) ion beams impacting on water, tissue, and bone, taking into account nuclear fragmentation processes. Geant4 is a C++ code that uses object-oriented approaches to simulate particle movement in matter.

It is a toolbox that uses the Monte Carlo approach to simulate particle passage through materials. The Geant4 toolbox can simulate a wide variety of physical processes, such as electromagnetic, hadronic, and optical processes. It offers methods for handling setup geometry, tracking main and secondary particles, and estimating their energy deposition in matter, as well as fundamental simulation operations. The setup for hadron therapy was meticulously modeled. A thorough simulation of the passive beamline was performed. Furthermore, the materials utilized were checked using a custom Geant4 code to ensure that their compositions and densities were correct.

Robert Wilson was the first to recognize the therapeutic benefits of particle radiation in the 1940s. Particle therapy has grown rapidly since then, with centers all around the world treating patients with protons and other heavy ions, including carbon ions. In 1994, the National Institute of Radiologic Sciences (NIRS) in Chiba, Japan, established the first heavy-ion accelerator for therapeutic usage.

Over 20,000 patients have been treated with carbon ion radiation therapy (CIRT) since then. There are currently five nations and 13 treatment centers using CIRT. Localized prostate cancer affects 22 % of patients treated with NIRS, with additional prevalent sites including bone and soft tissue which are 13 %, while head and neck are 11 %.

It is worth noting that as compared to photons or protons, they have greater linear energy transfer (LET) values. The energy transfer from a radiation beam to the medium it traverses per unit length is known as LET. Carbon beams with a higher LET have dramatically distinct biological effects at the DNA level.

The ratio of dose from a certain radiation modality needed to generate the same amount of tumor death as a reference dose, which is commonly X-rays of 250 kVp, is referred to as relative biological effectiveness (RBE). As a result, the RBE for photons is 1.

Although RBE is a complex entity that is usually dependent on the test radiation's LET, physical dose, irradiated tumor type, depth of tumor, end point, and other factors, RBE for protons is generally accepted to be 1.1 (despite the fact that proton RBE increases toward the end of the range), whereas RBE for carbon ions has been generally accepted to be in the 2–3 range or higher.

The ability of carbon ions to generate more complicated DNA damage than photons or protons is directly related to their greater RBE. These complicated DNA damages are likely to overwhelm the cell's ability to repair itself, resulting in accelerated tumor death. Unlike photon radiotherapy, where cell kill is based on the cell cycle, CIRT's cell killing potential is independent of the cell cycle.

## 2.6 Air

The molecules of gases in atmosphere composed of nitrogen (78 %), oxygen (21 %), and argon (1 %). Other molecules can be found in the atmosphere, but in very minor quantities. The composition of the atmosphere does not change much through the lower layers. As at higher level, the air molecules become less existent.

Although the stratosphere is generally composed of the same chemical composition, there is a difference in its chemical composition. The ozone layer is composed of various concentrations of ozone molecules. The ozone molecules are composed of three oxygen atoms. They keep the Sun's intense rays at bay. Scientists and researchers at the National Center for Atmospheric Research are monitoring the thin layer at the South Pole, which they refer to as a hole.

The composition changes as gases are ionized in the thermosphere. While still dominated by oxygen and nitrogen, the bonds between these atoms are broken. The exosphere is the region below Earth's atmosphere where air molecules can escape.

Air can be found anywhere on Earth. It can be found in the soil and the surface layer of the Earth. Earth's atmosphere is composed of air. This layer is divided up into different sections depending on the height and temperature of the atmosphere.

The first layer of air that is located near the Earth's surface is called the troposphere. This layer height 11 in kilometers. When moving in the troposphere, the temperature drops significantly. The weather on Earth is mainly determined by the conditions within the troposphere. The upper layer of the atmosphere is called the tropopause. It is located around eight to ten kilometers above the Earth's surface. At the equator, the tropopause layer is located about 17–18 km above Earth.

The stratosphere is the second layer of air that is located above the stratosphere. The temperature drops below the surface about $-55$ °C. The higher the temperature, the more oxygen is produced, and the less is absorbed by the Earth's atmosphere. In the stratosphere, solar radiation causes the $O_3$ and ozone to form and reach the Earth's surface at altitudes of 20–40 km. The upper part of the stratosphere, which is referred to as stratopauze, is caused by a reaction known as ozonesphere.

The third layer of air known as the mesosphere is located almost 52 km above the Earth's surface. The lower part of the mesosphere, which is known as the mesopause, is experiencing a cooling trend.

The fourth layer of air is known as the thermosphere. It is located over 90 km above Earth's surface. The intense heat produced by this layer causes it to reach a temperature of over 1000 °C. The density of air is low in this layer, which makes it hard for molecules to get separated. The lighter molecules can escape through the exosphere, which is the layer below the thermosphere. This is because the exosphere has a border that fades into space.

The lowest part of the atmosphere is usually referred to as the hemisphere, as its air composition is fairly constant. The air around this region is referred to as hydrosphere. Air molecules are constantly moving. They undergo constant thermal motion. The simplest type is uniform translation. It can be translated at a speed of 500 ms$^{-1}$. A molecule has three degrees of movement which are up and down, left and backward, and forward. Air molecules are only 0.6% of the volume they occupy. This means that the air is a very sparse gas with a vacuum atmosphere.

There are $2.7 \times 10^{19}$ atoms in air molecules. This high number density and the large translational speeds imply that the molecules are constantly colliding. The average molecule undergoes five collisions every nanosecond. One nanosecond corresponds to the time it takes light to travel 30 cm in a vacuum.

Other forms of random thermal motions can be observed in the molecules depending on their shape. Monatomic gases can only translate. Diatomic gases like oxygen and nitrogen can also undergo rotation and vibration. The two degrees of freedom are degenerated in the sense that they cannot be distinguished between them. The concept of the single vibrational mode allows energy to be stored in its kinetic energy.

## 2.7 Hydrogen

Hydrogen is a non-volatile liquid substance that can be colorless, tasteless, and odorless. It is the simplest member of the chemical family. The hydrogen atom has a nucleus composed of a proton bearing unit and an electron, which are also associated with this nucleus. Under normal conditions, hydrogen gas is composed of a pair of hydrogen atoms, which are $H_2$. Hydrogen is a substance that can transform into water by burning with oxygen. This chemical feature is the reason why it got its name.

Although hydrogen is the most widely used element in the universe, it only makes up about 0.14 % of Earth's crust. This phenomenon occurs in vast quantities in different parts of the world's water bodies. Hydrogen is a component of carbon compounds. It is present in all animal tissues and petroleum. Even though it is widely believed that there are more compounds of carbon than any other element, the reality is that there are more compounds of hydrogen than of carbon. Elementary hydrogen is a major component of the industrial chemical group ammonia. It is used in the manufacture of ammonia and in the hydrogenation of organic compounds.

There are three known isotopes of hydrogen. The mass numbers of these are 1, 2, and 3, respectively. The most abundant is the mass 1 isotope, hydrogen. The mass 2 isotope, which is deuterium, which has a nucleus composed of one proton and one neutron, is 0.0156 percent hydrogen. Tritium is the mass 3 isotope that constitutes about 10–16 % of hydrogen. The practice of naming hydrogen isotopes is justified due to the differences in their properties.

During the sixteenth century, Paracelsus unknowingly experimented with hydrogen after discovering that a flammable gas could be created when a metal was mixed with acid. Henry Cavendish, a physicist and chemist, discovered in 1766 explained that hydrogen was different from other flammable gases because its density and the amount of its chemical composition were different.

In 1781, Charles Cavendish proved that water was formed when the hydrogen was burned. In 1929, Paul Hartck and Karl Friedrich Bonhoeffer showed that ordinary hydrogen is composed of two kinds of molecules which are para-hydrogen and ortho-hydrogen. Due to the simple structure of hydrogen, its properties can be easily calculated. And due to its complexity, hydrogen is often used as a model for other atoms. These results are also applied to other atoms.

The properties of molecular hydrogen are discussed in Tables 2.1 and 2.2, respectively. The melting and boiling points are respectively low and powerful forces of attraction. The intermolecular forces that keep the temperature of hydrogen gas at room temperature also known to exist. This phenomenon shows that when hydrogen gas expands, its temperature rises.

According to the thermodynamic principles, if the forces between hydrogen molecules are equal to the attractive forces between them, then the expansion would

**Table 2.1** Atomic properties of normal hydrogen and deuterium

| Atomic hydrogen | | |
| --- | --- | --- |
| | Normal hydrogen | Deuterium |
| Atomic number | 1 | 1 |
| Atomic weight | 1.0080 | 2.0141 |
| Ionization potential (eV) | 13.595 | 13.600 |
| Electron affinity (eV) | 0.7542 | 0.754 |
| Nuclear spin | 1/2 | 1 |
| Nuclear magnetic moment (nuclear magnetons) | 2.7927 | 0.8574 |
| Nuclear quadrupole moment | 0 | $2.77 \times 10^{-27}$ cm$^2$ |
| Electronegativity (Pauling) | 2.1 | ~ 2.1 |

**Table 2.2** Molecular properties of normal hydrogen and deuterium

| Molecular hydrogen | | |
| --- | --- | --- |
| | Normal hydrogen | Deuterium |
| Bond distance (Å) | 0.7416 | 0.7416 |
| Dissociation energy (25 °C) | 104.19 kilocalories per mole | 105.97 kilocalories per mole |
| Ionazation potential (eV) | 15.427 | 15.457 |
| Density of solidg cm$^{-3}$ | 0.08671 | 0.1967 |
| Melting point°C | − 259.20 | − 254.43 |
| Heat of fusion | 28 cal per mole | 47 cal per mole |
| Density of liquid | 0.07099 (− 252.78°) | 0.1630 (− 249.75°) |
| Boiling point (°C) | − 252.77 | − 249.49 |
| Heat of vaporization | 216 cal per mole | 293 cal per mole |
| Critical temperature°C | − 240.0 | − 243.8 |
| Critical pressure | 13.0 atmospheres | 16.4 atmospheres |
| Critical density (g cm$^{-3}$) | 0.0310 | 0.0668 |
| Heat of combustion to water (g) | − 57.796 kilocalories per mole | − 59.564 kilocalories per mole |

cause the hydrogen to expand. In addition, as the attractive forces predominate, hydrogen, which is the main component, cools upon being permitted to expand below temperature of $-68\ ^{\circ}C$. The cooling effect is caused by the temperature difference between liquid nitrogen and hydrogen gas. This effect can be achieved by reducing the liquid nitrogen to a certain temperature.

Hydrogen is transparent to visible light, to infrared light, and to ultraviolet light to wavelengths below $1800\ \mathring{A}$. This gas has a velocity higher than that of any other gas at certain temperature. It diffuses much faster than any other gas. As a result, hydrogen is the fastest-moving gas in the world. Its kinetic energy is distributed much faster than that of other gases. A hydrogen molecule is composed of two protons and two electrons held together by electrostatic forces. It can exist in a number of different energy levels.

These hydrides are often referred to as covalently bonded. When a hydrogen atom is bound to two different electronegative atoms, it becomes hydrogen bonded. The strongest hydrogen bonds are formed by the highly electronegative atoms of fluorine, oxygen, and nitrogen. The hydrogen atom is a link between two fluorine atoms.

This connection is known as a bifluoride ion, or HF2. When ice melts, it breaks the hydrogen bonds on the structure, and as the result, the collapsing between structure occurs when the density increases.

Hydrogen bonding is a process utilized in biology to determine the configurations of molecules. It is important because it plays a major role in determining the orientation of chains. Extensive hydrogen bonding explains why some of the heavier analogues of hydrogen, such as hydrogen fluoride and water, have higher boiling points than their lighter counterparts. This thermal energy is required to break up the bonds between the hydrogen atoms and allow the vaporization to happen.

The hydrogen in a strong acid such as hydrochloric or nitric acts differently than that in a non-strong acid. When these acids dissolve in water, the resulting hydrogen, $H^+$, is separated from the anion, and the interaction between water molecules is occurred. The proton is attached to a water molecule to form an ion, which then bonds to other water molecules.

The reduction of $H^+$ is a chemical reaction that occurs when an atomic ion gains one or more electrons. The energy needed to generate this reaction is expressed as a reduction potential. The reduction potential of hydrogen is zero, and all metals have negative reduction potentials. In principle, dilute hydrogen with a strong acid solution displaces it. Positive reduction potentials are metals that are inert toward the hydrogen ion.

## 2.8 Laser-Induced Carbon Plasma

The laser ablation is widely used to obtain a variety of carbon-related materials, such as diamond-like carbon, fullerenes, and carbon nanotubes. The evaporation of a material is strongly affected by plasma formation. The dense plasma absorbs energy from the laser beam, and its temperature and pressure grow.

The thickness of the plasma layer is small compared with its other dimensions; therefore, the pressure gradient inside the major part of this layer is large and nearly perpendicular to the surface. Such a pressure gradient accelerates the plume to a high velocity perpendicular to the target. The hydrodynamic model which describes the target heating, formation of the plasma, and its expansion consists of equations of conservation of mass, momentum, and energy and is solved with the use of the Fluent software package.

It is assumed that the carbon plume expands to ambient air at a pressure of $10^{-3}$ Pa. It is also assumed that the electron temperature $T_e$ at the end of the Knudsen layer equals the target surface temperature $T_s$ contrary to the temperature of the heavy particles $T_h$ which, according to the theory of the Knudsen layer, is 0.67 $T_s$. The estimation shows that there is no time for energy equilibration between the carbon atoms and the electrons in the Knudsen layer. Hence, it is assumed that the vapor is in the Saha equilibrium at a temperature $T_e = T_s$ and $T_e/T_h = 1.5$.

Since the energy of the laser beam is supplied to electrons, the electron temperature will always exceed the temperature of heavy particles during the laser pulse. Therefore, the temperature ratio $T_e/T_h = 1.5$ is kept for first 9 ns of calculations. Then the temperature reaches 25 kK and the electron density $N_e \approx 1 \times 1026$ m$^{-3}$, and $T_e/T_h$ tends to unity.

After the cessation of the laser pulse, the energy equilibration time between electrons and heavy particles is a few nanoseconds. The energy source term IL was used in the form which fits the shape of the laser pulse, and the plasma absorption coefficient included all possible absorption mechanisms.

The same total absorption coefficient as in supplemented with the Mie absorption has been used. Radiative losses included total continuum and line radiation but were treated in a simplified way. It was assumed that at pressures higher than $0.5 \times 10^5$ Pa, the line radiation with $\lambda < 200$ nm was trapped in the plasma and hence was not included in the energy loss function. At lower pressures, this radiation was gradually included into energy losses.

The behaviors of carbon plasma have been studied from the second half of the nineteenth century, in the form of astronomical observations of comet spectra. Modern research on carbon plasmas may help in understanding fullerene and carbon nanotube production processes, their kinetics and molecular mechanisms, and the way in which carbon clusters and subsequent forms of condensed phase carbon nanostructures, in particular thin films, are formed.

As theory and experiment move forward in a synergistic and complementary way, electronic and vibrational properties probed with modern computer simulation methods promise insights into carbon dynamics. To some extent, the laser-induced breakdown spectroscopy (LIBS) technique owes its widespread use to these carbon plasma applications.

The LIBS method may be used to generate plasmas from any sample forms and phases. The emission spectra of these plasmas can be recorded with fast time resolution, and these spectra are then used to study the time evolution of plasma composition and temperature at nano- or picosecond resolution levels. The LIBS method has been named this way due to the fact that laser photons ionize the media, and in these ionized media volume electric discharges occur via dielectric breakdown processes.

The components of the plasma are usually not in thermodynamic equilibrium and are usually highly excited. The energy required to break chemical bonds and ionize atoms is provided by the thermal and optical energy from the absorption and scattering of laser photons. We have utilized the molecular emission Swan band of the C2 radical in the carbon plasma for vibrational–rotational plasma temperature measurements.

Swan bands are a characteristic of the spectra of carbon stars, comets, and of burning hydrocarbon fuels. They are named after the Scottish physicist William Swan (1818–1894) who first studied the spectral analysis of C2 in 1856. This band corresponds to a triplet–triplet electronic transition. The spectroscopy of the C2 radical is very extensive.

A stationary graphite target has been used for plasma generation, and a 10 Hz flash lamp pumped Nd:YAG laser for generating the plasma. The laser beam was narrowed using a two-lens inverted telescope to increase laser fluence on the sample surface. By using the 1064 nm laser radiation with telescopic beam narrowing and electronic pulse energy reduction, the typical fluence was about 150 MW/cm$^2$, while using the 532 nm laser radiation, this figure was around 60 MW/cm$^2$.

The graphite sample was placed inside a cylindrical quartz plasma cuvette that could be evacuated and filled with a background gas. In these experiments, helium has been used as background gas at 50 Torr pressure. The emitted plasma light was collected by a multi-component optical fiber perpendicularly to the cuvette axis. On the receiving end of the fiber, a small quartz lens is collected light from a wide angle.

At the end of the fiber bundle, the individual fibers were arranged in a plane to fit the entrance slit of the spectrograph. Two gratings were used, a lower resolution one with 150 lines/mm and blazing angle at 500 nm and a 1200 lines/mm grating blazed at 300 nm. Scattered laser light from the 532 nm laser radiation was blocked by a notch filter. Occasionally, light below 300 nm was also filtered out by a high pass filter, removing light below 300 nm. In this way, second-order spectral lines arising from the UV region were removed.

Emission spectra were observed using a triple grating spectrograph of Czerny-Turner arrangement and a focal length of 303 mm, equipped with a water-cooled (− 35 °C) intensified Andor IStar DH 720-18U-03 ICCD detector. An electronic gain factor of 50 was generally employed (the maximum gain factor is 255) and an entrance slitwidth of 50 μm. Ten spectra were accumulated to increase signal to noise ratio. More extensive accumulation would have resulted in fast blackening of the cuvette walls.

Occasionally, these parameters were changed to optimize the intensity and signal to noise of the spectra and also to minimize blackening of the LIBS cuvette walls by the ablated material from the graphite target. Emission spectra were wavelength calibrated either by using the built-in automatic wavelength calibration function or by using a Pen-Ray low-pressure cold cathode mercury-argon lamp and processed by the Andor proprietary Solis software. The automatic calibration is faster but less accurate than using the calibration lamp.

The emission spectra of the plasma plume were registered with the use of a spectrograph/monochromator and an ICCD camera. The camera was gated by a digital delay generator triggered by the signal from the laser. The exposure time (gate width) was 10 ns (ns). The delay time changed from 15 to 180 ns. Plasma temperature and electron density in the early phase of expansion were determined using optical emission spectroscopy.

The electron temperature and density were determined from the intensities and profiles of C IV, C III, and C II lines. The expansion velocity of the plasma plume was determined using the time-of-flight method as shown in Fig. 2.2. Temporal evolution of the intensity of specific spectral lines was measured with a monochromator and photomultiplier at various distances from the target and registered on an oscilloscope. These measurements allowed us to determine the expansion velocity of the plasma plume using the time-of-flight method.

The formation of laser-induced carbon plasma includes energy transfer, evaporation of solid material, and generation of dense plasma. Diamond-like carbon (DLC)

**Fig. 2.2** Time-of-flight method

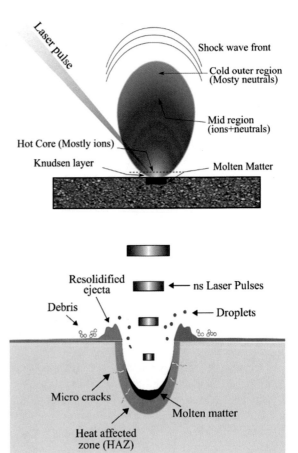

films at moderate laser irradiances of about $10^{10}$ W cm$^{-2}$ with 1.06 µm pulsed laser deposition have been investigated. Maximum energy of 300 mJ, 10 Hz pulse repetition frequency, and 9 ns pulse width have been used in order to produce carbon plasma. Helium gas has been filled in plasma chamber for a specific pressure to help in the cooling process.

The vibrational temperature of $C_2$ was identified, and it was found that the degree of ionization and vibrational of the temperature of $C_2$ species in plasma was increased with increase in laser irradiance. The characteristics of the spectral emission intensity from $C_2$ toward the distance from graphite target have been determined under different ambient helium gas pressures.

Laser ablation of ion-irradiated CR-39 has been investigated in 2006. The change in the track registration properties of CR-39 after irradiation with multiple pulses of $CO_2$ laser has been examined, and it is observed that after irradiated with $CO_2$ laser argon, cadmium and air molecular ions ($N_2$ and $O_2$, etc.) were generated using energy of 300 keV proton beam.

The diffraction patterns, circular fringes, holes drilling, and droplets have been observed on the surface of CR-39. For these purposes, an amount of 6.25 M of NaOH solution was used for the etching process with temperature at 70 °C for 6 h to measure the track diameters and track density of ions. The circular shape of the track was changed to elliptical, hexagonal, and triangular shapes. This is proven that registered tracks can behave as the defects of polymer.

Modern research on carbon plasmas has been conducted in the context of astronomical observations. These studies provide insight into the behavior of carbon plasmas and their various production processes. Through the use of computer simulations, scientists are able to study the interactions among various vibrational properties. To some extent, the laser-induced breakdown spectroscopy (LIBS) technique owes its widespread use to these carbon plasma applications.

## 2.9  CR-39

The solid-state nuclear track detectors or CR-39 materials have been widely used for the detection and measurement of charged particles. The abbreviation of CR-39 stands for "Columbia Resin #39." It is because it was the 39th formula of a thermosetting plastic developed by the Columbia Resins project in 1940. CR-39 is a plastic polymer named allyl diglycol carbonate (ADC) with chemical composition $C_{12}H_{18}O_7$ and acts as a solid-state nuclear track detector.

The polymerization schedule of ADC monomers using IPP is generally 20 h long with a maximum temperature of 95 °C. The elevated temperatures can be supplied using a water bath or a forced air oven. In the mean times, benzoyl peroxide (BPO) is an alternative organic peroxide that may be used to polymerize ADC. Pure BPO is crystalline and less volatile than diisopropyl peroxydicarbonate. By using BPO results in a polymer that has a higher yellowness index, the peroxide takes longer to dissolve into ADC at room temperature.

The solid-state nuclear track detectors is an important and useful technique in order to detect charged particles with energy range of 10 keV up to few hundreds MeV. The CR-39 detector is suitable for protons, deuterons, and alpha detection because of their high sensitivity toward the particles. Energy of carbon ions is determined based on the laser ablation of matter by using pure graphite as a target. Due to the high energy of carbon ions, a shape of cone was formed by accelerating the carbon ions to the target surface.

CR-39 is one of the detectors that have been widely used for the purpose of detection and measurement of charged particles energy. The monomer structure of CR-39 is presented in Fig. 2.1. For etching process, CR-39 detectors were exposed to carbon ions with energies between 1 and 10 MeV of 6 M KOH concentration at (60 $\pm$ 2) °C over 12 h. The pit diameter was measured for every different etching time. It was hypothesized that the diameter of pit becomes larger as the time for etching process increases and the etched pits circles due to normal ion incidence.

In 2011, an experiment to characterize the platinum and gold ions from laser-produced plasma using CR-39 has been done. The experiment was carried out using Nd:YAG laser (1064 nm wavelength, 9–14 ns pulse width, and 1.1 MW energy) under vacuum pressure of ~ $10^{-3}$ torr.

The emitted ions track was observed on the SSNTD's surface. A 6.25 M of NaOH solution at temperature (70 $\pm$ 1) °C for 8 h was used for etching process in order to observe the tracks on the surface of SSNTD's. The energies of ion can be determined from the diameter of etched tracks. As the results, energies of 0.02–4.33 keV and 0.03–6.91 keV have been estimated for gold and platinum ions, respectively.

In 2014, CR-39 polymer track detector was irradiated with gamma ray to study the effect of heat treatment on the bulk etch rate, track density, and the damage in the sample surface for which the samples have been annealed at certain temperature (50, 100, and 150 °C). A 6.25 M NaOH concentration was heated for 6 h at (70 $\pm$ 2) °C for the etching process. The result showed that the effect of heat treatment to the gamma-irradiated CR-39 samples gives a serious degradation in the polymeric material.

### 2.9.1  Q-Switched Nd:YAG Laser

Nd:YAG or neodymium-doped yttrium aluminum garnet, Nd:$Y_3Al_5O_{12}$, is a crystal which is used as a lasing medium especially for solid-state lasers. The dopant, triply ionized neodymium (Nd (III)), typically replaces a small fraction (about 1 %) of the yttrium ions in the host crystal structure of the yttrium aluminum garnet (YAG), since the two ions are of similar size.

It is the neodymium ion which provides the lasing activity in the crystal in the same fashion as red chromium ion in ruby lasers. Nd:YAG laser is a low power consumption laser with high gain, good thermal properties, and has good mechanical properties. The efficiency of Nd:YAG laser is very high as compared with the ruby laser.

**Fig. 2.3** Q-switched Nd:YAG laser

Nd:YAG laser is optically pumped using a flashtube or laser diodes. Figure 2.3 shows the examples of Q-switched Nd:YAG laser. This laser is one of the most common types of lasers and is used for many different applications. Nd:YAG laser typically emits light with a wavelength of 1064 nm (in the infrared range). However, there are also transitions near 946, 1120, 1320, and 1440 nm wavelength as well.

Nd:YAG laser operates in both pulsed and continuous modes. Pulsed Nd:YAG laser is typically operated in the so-called Q-switching mode, which is an optical switch inserted in the laser cavity waiting for a maximum population inversion in the neodymium ions before it opens.

The light wave can run through the cavity, depopulating the excited laser medium at maximum population inversion. In this Q-switched mode, output powers of 250 MW and pulse durations of 10–25 ns have been achieved. The high-intensity pulses may be efficiently frequency doubled to generate laser light at 532 nm or higher harmonics at 355, 266, and 213 nm, respectively.

Nd:YAG laser absorbs mostly in the bands between 730 and 760 nm and 790 and 820 nm, respectively. At low current densities, krypton flash lamps have higher output in those bands than do the more common xenon lamps, which produce more light at around 900 nm. The amount of the neodymium dopant in the material varies according to its use. For continuous wave output, the doping is significantly lower than for pulsed lasers. The lightly doped CW rods can be optically distinguished by being less colored (almost white), while highly doped rods are pink-purplish.

Other common host materials for neodymium are yttrium lithium fluoride (YLF) with 1047 and 1053 nm wavelength, yttrium orthovanadate ($YVO_4$) at 1064 nm wavelength, and glass. A particular host material is chosen in order to obtain a desired combination of optical, mechanical, and thermal properties. Nd:YAG lasers and variants are pumped either by flashtubes, continuous gas discharge lamps, or near-infrared laser diodes (DPSS lasers).

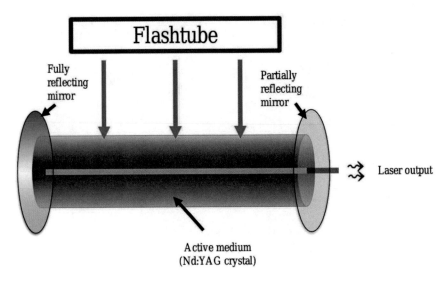

**Fig. 2.4** Operational principles of Nd:YAG laser

Figure 2.4 shows the schematic diagram of the operational principles of Nd:YAG laser. Nd:YAG laser consists of three important elements, which are energy source, active medium, and optical resonator. The energy source or pump source supplies energy to the active medium to achieve the population inversion of the laser. In Nd:YAG laser, light energy sources such as flashtube or laser diodes are used as energy source to supply energy to the active medium.

In the past, flashtubes are mostly used as pump source because of its low cost. But nowadays, laser diodes are preferred over flashtubes because of its high efficiency and low cost.

The active medium or laser medium of the Nd:YAG laser is made up of a synthetic crystalline material (yttrium aluminum garnet (YAG)) doped with a neodymium (Nd) chemical element. The lower energy-state electrons of the neodymium ions are excited to the higher energy state to provide lasing action in the active medium. The Nd:YAG crystal is placed between two mirrors.

These two mirrors are optically coated or silvered. Each mirror is coated or silvered differently. One mirror is fully silvered, whereas another mirror is partially silvered. The mirror which is fully silvered will completely reflect the light and is known as fully reflecting mirror. On the other hand, the mirror which is partially silvered will reflect most part of the light but allows a small portion of light through it to produce the laser beam. This mirror is known as a partially reflecting mirror.

Nd:YAG laser is a four-level laser system, which means that the four energy levels are involved in this laser action. The light energy sources such as flashtubes or laser diodes are commonly used to supply energy to the active medium. In Nd:YAG laser, the lower energy-state electrons in the neodymium ions are excited to the higher energy state to achieve the population inversion.

Consider a Nd:YAG crystal active medium consisting of four energy levels $E_1$, $E_2$, $E_3$, and $E_4$ with $N$ number of electrons, as illustrated in Fig. 2.5. The number of electrons in the energy states $E_1$, $E_2$, $E_3$, and $E_4$ will be $N_1$, $N_2$, $N_3$, and $N_4$, respectively. Assume that the energy levels will be $E_1 < E_2 < E_3 < E_4$.

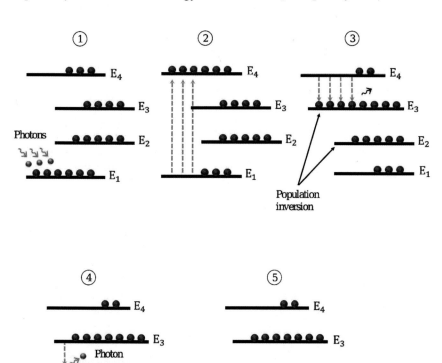

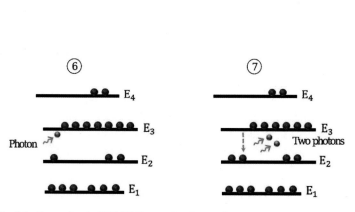

**Fig. 2.5** Energy level of Nd:YAG crystal active medium

The lifetime of pump state or higher energy state, $E_4$, is very small, which is about 230 μs. So, the electrons in the energy state $E_4$ do not stay for long time of period. After a short period, the electrons will fall into the next lower energy state or metastable state $E_3$ by releasing non-radiation energy without emitting photons.

The lifetime of metastable state $E_3$ is higher as compared to the lifetime of pump state $E_4$. Therefore, the electrons reach $E_3$ much faster than they leave $E_3$. These results can increase in the number of electrons in the metastable state $E_3$, and hence, the population inversion is achieved.

After some period, the electrons in the metastable state $E_3$ will fall into the next lower energy state $E_2$ by releasing photons or light. The emission of photons in this manner is called spontaneous emission. The lifetime of energy state $E_2$ is very small, just like the energy state at $E_4$. Therefore, after a short period, the electrons in the energy state $E_2$ will fall back to the ground state $E_1$ by releasing radiationless energy.

When the photon emitted due to the spontaneous emission is interacted with the other metastable state electron, it stimulates that electron and makes it fall into the lower energy state by releasing the photon. As a result, two photons are released, and the emission of photons in this manner is called stimulated emission of radiation.

When these two photons again interacted with the metastable state electrons, four photons are released. Likewise, millions of photons are emitted. Thus, optical gain is achieved. Spontaneous emission is a natural process, but stimulated emission is not a natural process. To achieve the stimulated emission, external photons or light should be supplied to the active medium.

The Nd:YAG active medium generates photons or light due to spontaneous emission. The light or photons generated in the active medium will bounce back and forth between the two mirrors. This stimulates other electrons to fall into the lower energy state by releasing photons or light. Likewise, millions of electrons are stimulated to emit photons. The light generated within the active medium is reflected many times between the mirrors before it escapes through the partially reflecting mirror.

### 2.9.2  Plasma

Plasma is one of the four fundamental states of matter and was first described by chemist Irving Langmuir in the 1920s. It consists of a gas of ions (atoms which have some of their orbital electrons remove) and free electrons. Plasma can be artificially generated by heating or subjecting a neutral gas to a strong electromagnetic field to the point where an ionized gaseous substance becomes increasingly electrically conductive. The resulting charged ions and electrons are influenced by long-range electromagnetic fields, making the plasma dynamics more sensitive to these fields than a neutral gas.

Plasma and ionized gases have properties and display behaviors unlike those of the other states, and the transition between them is mostly a matter of nomenclature and subject to interpretation. Based on the temperature and density of the environment that contains a plasma, partially ionized or fully ionized forms of plasma may be produced.

Neon signs and lightning are examples of partially ionized plasmas. The Earth's ionosphere is a plasma, and the magnetosphere contains plasma in the Earth's surrounding space environment. The interior of the Sun is an example of fully ionized plasma, along with the solar corona and stars.

Ions can be in the form of positive or negative charges. Positive charges in ions are achieved by stripping away electrons orbiting the atomic nuclei, where the total number of electrons removed is related to either increasing temperature or the local density of other ionized matter.

This can be accompanied by the dissociation of molecular bonds. This process is distinctly different from chemical processes of ion interactions in liquids or the behavior of shared ions in metals. The response of plasma to electromagnetic fields is used in many modern technological devices, such as plasma televisions or plasma etching.

Plasma may be the most abundant form of ordinary matter in the universe, although this hypothesis is currently tentative based on the existence and unknown properties of dark matter. Plasma is mostly associated with stars, extending to the rarefied intracluster medium and possibly the intergalactic regions. Plasma is a state of matter in which an ionized gaseous substance becomes highly electrically to the point that long-range electric and magnetic field dominate the behavior of the matter.

The plasma state can be contrasted with the other states, which are solid, liquid, and gas. Plasma is an electrically neutral medium of unbound positive and negative particles. Although these particles are unbound, they are not "free" in the sense of not experiencing forces. Moving charged particles generate an electric current within a magnetic field, and any movement of a charged plasma particle affects and is affected by the fields created by the other charges.

On top of that, there are three factors that define a plasma, which are the plasma approximation, bulk interactions, and plasma frequency. The plasma approximation applies when the plasma parameter, $\Lambda$, represents the number of charge carriers within a sphere, called the Debye sphere.

The radius of Debye sphere is the Debye screening length surrounding a given charged particle. It is sufficiently high as to shield the electrostatic influence of the particle outside of the sphere. The Debye screening length is short compared to the physical size of the plasma.

This criterion means that the interactions in the bulk of the plasma are more important than those at its edges, where boundary effects may take place. When this criterion is satisfied, the plasma is quasi-neutral. This is called as bulk interactions. For the plasma frequency, the electron plasma frequency is large compared to the electron-neutral collision frequency. When this condition is valid, electrostatic interactions dominate over the processes of ordinary gas kinetics.

### 2.9.3 Plasma Temperature

Plasma temperature is commonly measured in Kelvin or electron volts (eV) and is informally a measure of the thermal kinetic energy per particle. High temperatures are usually needed to sustain the ionization, which is a defining feature of a plasma. The degree of plasma ionization is determined by the electron temperature, relative to the ionization energy in a relationship called the Saha equation. At low temperatures, ions and electrons tend to recombine into bound-state atoms, and the plasma will eventually become a gas.

In most cases, the electrons are close enough to thermal equilibrium that their temperature is relatively well-defined. This is true even when there is a significant deviation from a Maxwellian energy distribution function. For example, due to the UV radiation, energetic particles, or strong electric fields.

Because of the large difference in mass, the electrons come to thermodynamic equilibrium among themselves much faster than they come into equilibrium with the ions or neutral atoms. For this reason, the ion temperature may be very different from the electron temperature. This is common especially in weakly ionized technological plasmas, where the ions are often near the ambient temperature.

Based on the relative temperatures of the electrons, ions, and neutrals, plasmas are classified as thermal or non-thermal (also referred to as cold plasmas). Thermal plasmas have electrons and the heavy particles at the same temperature, while non-thermal plasmas, on the other hand, are non-equilibrium ionized gases, with two different temperatures, whereas electrons are much hotter.

A kind of common non-thermal plasma is the mercury-vapor gas within a fluorescent lamp, where the electrons gas reaches a temperature of 10,000 K, while the rest of the gas stays barely above room temperature. So, the bulb can even be touched with hands while operating. A particular and unusual case of inverse non-thermal plasma is the very high-temperature plasma produced by the Z machine, where ions are much hotter than electrons.

For plasma to exist, ionization is necessary. The term plasma density itself usually refers to the electron density. Electron density is the number of free electrons per unit volume. The degree of ionization of a plasma is the proportion of atoms that have lost or gained electrons and is controlled by the electron and ion temperatures and electron-ion versus electron-neutral collision frequencies.

In a plasma, the electron-ion collision frequency is much greater than the electron-neutral collision frequency. Therefore, with a weak degree of ionization, the electron-ion collision frequency can equal the electron-neutral collision frequency. The term fully ionized gas introduced by Lyman Spitzer does not mean the degree of ionization is unity, but only that the plasma is in a Coulomb collision-dominated regime. On the other hand, a partially or weakly ionized gas means the plasma is not dominated by Coulomb collisions, and most of technological plasmas are weakly ionized gases.

## 2.9.4  Plasma Potential

Plasma is often called the fourth state of matter, after solid, liquid, and gas. It is a gas-like state of matter that contains a significant number of charged particles, such as ions and free electrons. The charged particles in a plasma interact strongly with each other and with external electromagnetic fields. This behavior leads to unique properties, such as conductivity, the ability to generate and respond to magnetic fields, and the formation of structures like double layers and plasma waves.

Plasma is found in many natural and artificial environments, from stars and lightning bolts to fluorescent light bulbs and plasma TVs. It is used in a variety of industrial applications, such as welding, cutting, and surface treatment, as well as in the production of semiconductors and other high-tech materials. Plasma research also has implications for fusion energy, space science, and medicine, among other fields.

Plasmas are excellent conductors of electricity, which means that electric potentials are significant in their behavior. The average potential in the space between charged particles, known as the plasma potential or space potential, is independent of the method used to measure it. When an electrode is inserted into a plasma, its potential is generally considerably lower than the plasma potential, due to a phenomenon called the Debye sheath.

The small electric fields resulting from the excellent conductivity of plasmas led to the concept of quasi-neutrality, which suggests that the density of negative charges is roughly equal to that of positive charges over large volumes of the plasma. However, at the Debye length scale, there can be charge imbalances.

In some cases, double layers are formed, and the charge separation can extend for several tens of Debye lengths. The magnitude of potentials and electric fields in such cases must be determined using methods other than just calculating the net charge density.

In astrophysical plasmas, Debye screening prevents electric fields from affecting the plasma directly over distances greater than the Debye length. However, charged particles in the plasma generate and are influenced by magnetic fields, resulting in highly complex behaviors such as the formation of plasma double layers, which separate charges over several tens of Debye lengths.

## 2.9.5  Magnetization

Plasma with strong enough magnetic field to influence the motion of the charged particles is said to be magnetized. A common quantitative criterion is that a particle on average completes at least one gyration around the magnetic field before making a collision. It is often the case that the electrons are magnetized while the ions are not. Magnetized plasmas are anisotropic, meaning that their properties in the direction parallel to the magnetic field are different from those perpendicular to it.

A plasma is a hot ionized gas consisting of approximately equal numbers of positively charged ions and negatively charged electrons. The characteristics of plasmas are significantly different from those of ordinary neutral gases so that plasmas are considered a distinct fourth state of matter. For example, because plasmas are made up of electrically charged particles, they are strongly influenced by electric and magnetic fields while neutral gases are not. An example of such influence is the trapping of energetic charged particles along geomagnetic field lines to form the Van Allen radiation belts.

In addition to externally imposed fields such as the Earth's magnetic field or the interplanetary magnetic field, the plasma is acted upon by electric and magnetic fields created within the plasma itself. This can be done through localized charge concentrations and electric currents results from the differential motion of the ions and electrons.

The forces exerted by these fields on the charged particles that make up the plasma act over long distances and impart to the particles' behavior a coherent, collective quality that neutral gases do not display. Despite the existence of localized charge concentrations and electric potentials, a plasma is electrically quasi-neutral. In aggregate, there are approximately equal numbers of positively and negatively charged particles distributed so that their charges cancel.

It is estimated that 99% of the matter in the observable universe is in the plasma state. Stars, stellar, extragalactic jets, and the interstellar medium are examples of astrophysical plasmas. In our solar system, the Sun, the interplanetary medium, the magnetospheres and/or ionospheres of the Earth and other planets, as well as the ionospheres of comets and certain planetary moons, all consist of plasmas.

The plasmas of interest to space physicists are extremely tenuous, with densities dramatically lower than those achieved in laboratory vacuums. The density of the best laboratory vacuum is about 10 billion particles per cubic centimeter. In comparison, the density of the densest magnetospheric plasma region, the inner plasmasphere, is only 1000 particles per cubic centimeter, while that of the plasma sheet is less than 1 particle per cubic centimeter.

The temperatures of space plasmas are very high, ranging from several thousand degrees Celsius in the plasmasphere to several million degrees Celsius in the ring current. While the temperatures of the cooler plasmas of the ionosphere and plasmasphere are typically given in degrees Kelvin, those of the hotter magnetospheric plasmas are more commonly expressed in terms of the average kinetic energies of their constituent particles measured in electron volts.

An electron volt (eV) is the energy unit that an electron acquires as it is accelerated through a potential difference of one volt and is equivalent to 11,600 degrees Kelvin. Magnetospheric plasmas are often characterized as being "cold" or "hot." Although these labels are quite subjective, they are widely used in the space physics literature.

As a rule of thumb, plasmas with temperatures less than about 100 eV are "cold," while those with temperatures ranging from 100 eV to 30 keV can be considered "hot." The particles with higher energies, such as those that populate the radiation belt, are termed as "energetic."

Plasma is a cloud of protons, neutrons, and electrons where all the electrons have come loose from their respective molecules and atoms, giving the plasma the ability to act as a whole rather than as a bunch of atoms. A plasma is more like a gas than any of the other states of matter because the atoms are not in constant contact with each other, but it behaves differently from a gas. It has what scientists call collective behavior. This means that the plasma can flow like a liquid or it can contain areas that are like clumps of atoms sticking together.

# Chapter 3
# Research Design

## 3.1 Introduction

This research presents a characterization of carbon ions in laser-induced plasma using solid-state nuclear track detectors and investigates the effect of varying ambient environments and pressures. The diameter of the ion tracks is measured using image analysis software to calculate the energy of the carbon ions. The experimental setup involves exposing solid-state nuclear track detectors to laser-induced plasma in a vacuum chamber, which allows for the control of ambient environment and pressure during the experiments.

Statistical analysis is performed on the collected data to determine significant differences in the energy of the carbon ions under different conditions. The methodology employed in this book provides a reliable means of characterizing carbon ions in laser-induced plasma, enhancing our understanding of their behavior in different environments and contributing to the development of plasma-based technologies.

## 3.2 Experiment Equipment

Science and experimentation can be both enjoyable and hazardous. Proper knowledge of how to handle chemicals and laboratory instruments is essential to prevent accidents. While curiosity is a desirable trait, it is crucial to have accurate information about the instruments before conducting experiments. Without the correct knowledge, experimentation can be challenging and hazardous.

It is essential to select appropriate laboratory equipment and acquire accurate information about its use before engaging with it. Laboratory equipment can have dangerous side effects, and it is critical to know how to handle it properly. Knowledge of operating laboratory instruments is essential for their safe and effective use.

It is crucial to ensure that all equipment used is in good condition and has been calibrated to ensure the safety and validity of the results. For example, the operational

© The Author(s), under exclusive license to Springer Nature Singapore Pte Ltd. 2023
S. Daud, *Carbon Nanotubes*, SpringerBriefs in Applied Sciences and Technology,
https://doi.org/10.1007/978-981-99-4962-5_3

principles of the vacuum chamber, graphite electrodes, power supply, solid-state track detector, and optical microscope used in this research will be explained and discussed in detail.

### 3.2.1 Vacuum Chamber

A vacuum chamber is an essential tool in various scientific and industrial applications, particularly in the field of nanotechnology. It provides a controlled environment with low-pressure conditions, enabling researchers to study and manipulate materials under vacuum.

In carbon nanotube experiments, vacuum chambers play a crucial role in creating a clean and controlled atmosphere, free from contaminants and interference. These chambers are typically constructed using materials such as stainless steel or aluminum, known for their compatibility with vacuum conditions and minimal outgassing.

To maintain the desired vacuum level, a pumping system is employed, ranging from mechanical pumps to advanced turbo molecular or cryogenic pumps. Pressure measurement devices, including Pirani gauges and ion gauges, allow precise monitoring of the chamber's pressure.

In addition to maintaining a vacuum, vacuum chambers often incorporate other features to facilitate experiments with carbon nanotubes. Electrical feedthroughs are used to establish electrical connections inside the chamber, enabling measurements and electrical manipulation of the nanotubes.

Temperature control mechanisms, such as heating elements or cryogenic cooling systems, may also be integrated to study the behavior of carbon nanotubes under different temperature conditions. Moreover, gas handling systems can be installed to introduce specific gases or control the gas environment within the chamber, allowing researchers to explore the effects of different gas environments on the properties of carbon nanotubes.

Furthermore, vacuum chambers often include safety features to ensure the well-being of researchers and the integrity of the experiments. Pressure relief valves are utilized to prevent overpressure situations, while interlocks and vacuum interlocks enhance operator safety and protect the chamber from accidental exposure to atmospheric conditions.

By providing a well-controlled and isolated environment, vacuum chambers enable researchers to investigate the unique properties and potential applications of carbon nanotubes with high precision and accuracy. These chambers serve as critical tools in advancing our understanding of carbon nanotubes and their role in various fields, including electronics, energy storage, and biomedical applications. Figure 3.1 shows the vacuum chamber used in this experiment.

Figure 3.2 depicts an external view of a state-of-the-art vacuum chamber, a remarkable piece of scientific equipment designed to facilitate advanced research in

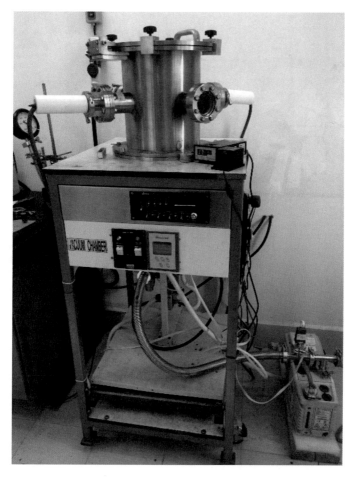

**Fig. 3.1** Vacuum chamber

various fields. The chamber stands as a robust and meticulously engineered structure, showcasing a sleek and modern design.

Its exterior is composed of high-quality materials, such as stainless steel or aluminum, known for their durability and compatibility with vacuum conditions. The smooth surfaces of the chamber reflect the ambient light, providing a glimpse of the precision and attention to detail involved in its construction.

From an external perspective, the figure reveals the intricate components and features that make the vacuum chamber a versatile tool for scientific exploration. It showcases the presence of carefully positioned ports, flanges, and fittings that serve as access points for various functionalities. These include translation stage control, camera, and pressure gauge, which allow for electrical connections, controlled gas environments, and the maintenance of the desired vacuum level, respectively. The

**Fig. 3.2** External view of the vacuum chamber

figure also highlights safety features, such as lock power supply which ensure the well-being of researchers and the protection of the chamber during experiments.

This external view of the vacuum chamber provides a glimpse into the world of cutting-edge research, where scientists and engineers harness the power of controlled vacuum environments to unlock the mysteries of carbon nanotubes and delve into the forefront of scientific discovery.

The internal view of the vacuum chamber shown in Fig. 3.3 offers a captivating glimpse into a realm of scientific exploration where precision and control converge. As the system peels back the metaphorical curtain, a meticulously crafted environment is revealed. The inner chamber, constructed with high-grade materials, emanates a sense of pristine purity. The walls, often composed of stainless steel or aluminum, are flawlessly smooth and free from imperfections to minimize any interaction with the experimental setup.

**Fig. 3.3** Setup inside vacuum chamber

Within the chamber, the focal point of the internal view, one can observe the carefully positioned sample holder or stage that cradles the carbon nanotube samples. This specialized component allows for precise manipulation and positioning of the nanotubes during experiments. Electrical feedthroughs, designed to maintain the vacuum integrity, elegantly pierce the chamber walls, providing essential connections for electrical measurements and controls.

The vacuum environment inside the chamber is meticulously maintained by an array of pumping systems. These pumps, strategically placed outside the field of view, efficiently evacuate the chamber to create and sustain the desired low-pressure conditions. Pressure measurement devices, such as Pirani gauges or ion gauges, seamlessly integrate into the chamber walls, ensuring accurate monitoring and control of the internal pressure.

The atmosphere within the chamber is carefully managed and controlled. Gas handling systems, located discreetly within the chamber or connected through specialized ports, facilitate the introduction of specific gases or the removal of impurities. This capability allows researchers to investigate the influence of different gas environments on the behavior and properties of carbon nanotubes.

As the figure portrays the internal view of the vacuum chamber, it provides a glimpse into the realm where scientific discovery unfolds. Within this controlled and isolated environment, scientists delve into the intricate world of carbon nanotubes, exploring their unique properties, unraveling their potential applications, and pushing the boundaries of human knowledge.

## 3.2.2 Graphite Electrode

Carbon graphite is one of the softest, hardest, and has diamond-like features materials found in nature. Graphite is an allotropic form of carbon. It is a substance that has soft grayish black crystalline properties. Due to these reasons, graphite was used in this research to produce arc plasma by arc discharge process. The example of carbon graphite used in this research is shown in Fig. 3.4.

The carbon atoms are arranged in flat plane of hexagonal ring, which are stack on each other. Each carbon atom is attached with four other carbon atoms, three neighboring atoms in same plane, and one carbon atom from other plan. Graphite is a good conductor of electricity because three out of four valance electrons are tightly bonded, while the fourth valance electron is loosely bonded between the planes. It has high melting point, which is above 3000 °C and stable over a wide range of temperatures.

In an electric arc furnace, where scrap from old automobiles or appliances is melted to make new steel, graphite electrodes serve as the primary heating element. Electric arc furnaces are less expensive to construct, compared with the traditional blast furnaces, which produce steel from iron ore and use coking coal as fuel. However, because they use steel waste and are driven by electricity, the cost of steelmaking is higher.

**Fig. 3.4** Graphite electrode

The electrodes are arranged into columns and form part of the furnace lid. The scrap steel is subsequently melted by an arc of tremendous heat created by electricity passing between the electrodes. Electrodes can be as small as 0.75 cm in diameter and as large as 2.8 m (9 ft) in length. The heaviest electrodes are above two metric tons, and one ton of steel requires up to 3 kg (6.6 lb) of graphite electrodes.

The electrode's tip will heat up to 3000 °C, which is half the temperature of the Sun's surface. Graphite is used for electrodes since it is the only material that can sustain such high temperatures. The molten steel is then poured into enormous buckets known as ladles when the furnace is tipped on its side. The molten steel is subsequently transported to the steel mill's caster, which transforms the scrap into new goods.

The electricity required for this process is enough to power a 100,000-person town. In a modern electric arc furnace, each melt takes around 90 min and produces 150 tons of steel, which is enough to construct 125 automobiles. The electrodes' primary raw material is needling coke, which can take up to six months to create due to operations such as baking and rebaking to convert the coke to graphite.

There are two types of needle coke, which are petroleum-based needle coke and coal-based needle coke. Both of these can be used to make graphite electrodes. Petroleum-based coke is a by-product of the oil refining process, whereas coal-based needle coke is generated from coal tar that is produced during the coke-making process.

The fundamental rationale for using graphite electrodes in electrolysis is that graphite is a good conductor. The structure of graphite allows for a high number of electrons to freely travel across many layers of atoms. Graphite bonds are formed of only three out of the four electron shells of the carbon atom, leaving the fourth electron to move freely.

These electrons operate as a strong conductor, allowing the electrolysis process to proceed without interruption. Graphite is also cost-effective, stable at high temperatures, and long-lasting. Due to these reasons, graphite electrodes are commonly used in electrolysis industry.

The reasons why graphite is used as an electrode in electrolysis. Because of the atomic structure of graphite, a high number of electrons are not bound, allowing them to move between graphite layers. Graphite's excellent conductive qualities are due to its enormous quantity of free electrons (electron delocalization). In addition to being an excellent conductor, graphite is also inexpensive, durable, and easy to get, the factors which contribute to its widespread use as an electrode.

Electrolysis uses graphite rods as electrodes because graphite's structure makes it a great conductor. Electricity can travel through graphite quickly due to the large number of delocalized electrons. Graphite is a most easier material to mound into a rod shape, cost-effective, and durable material.

Graphite electrode is very suitable for electrolysis because of its outstanding conductive qualities, high melting point (which allows it to be utilized in a wide range of electrolysis processes), and low cost, and graphite is a good choice for an electrolysis electrode. Positively charged ions (metals and hydrogen) can receive electrons from a negatively charged electrode due to graphite. Negatively charged ions, on the other hand, lose their electrons (oxidation).

### 3.2.3 Power Supply

A power supply is a device used to convert electrical energy from a source to an output that can be used to power various devices or systems. In other words, it is an electric power converter that takes in electrical energy and transforms it into a form that can be used by other electrical components.

In many applications, such as in the welding industry, a rectifier is an essential component of a power supply. A rectifier is an electrical device that converts alternating current (AC) to direct current (DC), which flows in only one direction. This process is crucial for many applications where a constant and reliable source of DC power is required.

The DC rectifier used in this experiment is a crucial component in the welding process as it provides a consistent and stable source of DC power to the electrodes. The device is equipped with an integrated circuit (IC) electronic collective printed circuit with constant voltage and current. This means that the output current and voltage are controlled by effectuating a stepless adjustment on the device, which allows for precise control of the power supply output.

The constant voltage and current provided by the DC rectifier are essential in welding as they ensure that the electrodes receive a consistent and stable source of power. This consistency is necessary for maintaining the quality of the weld and preventing any defects or inconsistencies that may arise from fluctuations in the power supply output.

Overall, the DC rectifier used in welding power supplies plays a critical role in ensuring a reliable and consistent power supply to the electrodes. With its integrated circuit technology and precise control of output current and voltage, the DC rectifier provides a reliable and stable source of power for a wide range of welding applications.

### 3.2.4 Solid-State Nuclear Track Detector

Solid-state nuclear track detector (SSNTD) is radiation detectors that are widely used in various fields of nuclear and particle physics, as well as in many other areas such as geology, environmental science, and medicine. They are passive detectors that are designed to detect and record the tracks of energetic particles produced by ionizing radiation in a thin layer of solid-state material.

The basic working principle of an SSNTD is based on the fact that energetic particles passing through a solid material produce ionization along their path, leaving behind a trail of charged particles. This ionization can cause permanent damage or displacement of atoms in the material, which can be visualized and analyzed by chemical etching or other imaging techniques.

The most common material used for SSNTDs is a polycarbonate film, which is an inexpensive and readily available material. The film is exposed to the ionizing radiation, and then it is chemically etched to reveal the tracks left by the particles. The tracks can be easily analyzed and counted using optical or scanning electron microscopy.

SSNTDs have many advantages over other types of radiation detectors. They are highly sensitive and can detect a wide range of ionizing radiation, from alpha particles to heavy ions. They are also highly durable, can withstand harsh environmental conditions, and do not require any power source or readout electronics, making them very practical for remote or long-term monitoring.

SSNTDs have many applications in radiation measurement and dosimetry. They are used to measure the radiation dose in various fields, such as medical radiology, nuclear power plants, and space exploration. They are also used in geology to study the age and history of rocks and minerals and in environmental science to monitor radon levels in homes and buildings.

In addition to their practical applications, SSNTDs are also used in fundamental research in nuclear and particle physics. They are used to study the properties of cosmic rays and high-energy particles produced by the Sun and other celestial sources. They are also used to study the interactions of particles with matter, which are crucial for understanding the behavior of matter at the atomic and subatomic level.

Overall, SSNTDs are powerful tools that have found many applications in science and technology. They are simple, reliable, and versatile and have contributed to many important discoveries and advancements in various fields of research.

### 3.2.5 Optical Microscope

The Nikon MICROPHOT-FXL optical microscope shown in Fig. 3.5 is a widely used instrument in scientific research and industrial applications due to its reliability and versatility. It is particularly favored for its compatibility with computer systems, which ensures efficient and accurate data collection and analysis.

One of the standout features of the MICROPHOT-FXL microscope is its range of powerful magnification levels, which span from $40\times$ to $2000\times$. This makes it an ideal tool for examining specimens of varying sizes and types. The microscope's solid construction and high-quality binocular sliding head add to its durability and dependability, while the 4 achromatic objectives and 2 pairs of wide field eyepieces ensure that specimens are displayed clearly and crisply.

In addition to its optics, the microscope is equipped with a solid mechanical stage that allows for precise movement and positioning of specimens. This is particularly important when studying ion tracks on CR-39 surfaces, as it allows for accurate measurement and analysis of these tracks.

The microscope's adjustable electrical illumination system is another noteworthy feature. Proper illumination is essential for accurate observation and data collection,

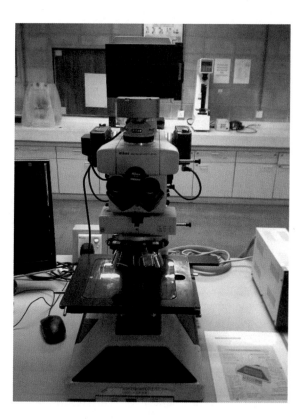

**Fig. 3.5** Optical microscope

and this system allows for the ideal level of illumination to be achieved. This can be adjusted according to different lighting conditions, enabling the examination of the specimen's textures and layers.

The MICROPHOT-FXL microscope is used extensively in various scientific fields, including microbiology, histology, and material science. Its ability to provide high-resolution images and data makes it an essential tool for researchers, scientists, and engineers.

The Nikon MICROPHOT-FXL optical microscope is a powerful and reliable instrument for studying ion tracks on CR-39 surfaces. Its high-quality optics, solid construction, and adjustable illumination system make it an ideal choice for researchers in a variety of fields, including materials science, nuclear physics, and geology.

### 3.2.6  Laser-Induced Carbon Plasma

The laser ablation is widely used to obtain a variety of carbon-related materials, such as diamond-like carbon, fullerenes, and carbon nanotubes. The evaporation of a material is strongly affected by plasma formation. The dense plasma absorbs energy from the laser beam which leads to its temperature and pressure grown.

The thickness of the plasma layer is small enough compared with its other dimensions. Therefore, the pressure gradient inside the major part of this layer is large and nearly perpendicular to the surface. Such pressure gradient accelerates the plume to a high velocity perpendicular to the target.

The hydrodynamic model which describes the target heating, formation of the plasma, and its expansion consists of equations of conservation of mass, momentum, and energy and is solved with the use of the Fluent software package. It is assumed that the carbon plume expands to ambient air at a pressure of $10^{-3}$ Pa.

It is also assumed that the electron temperature, $T_e$, at the end of the Knudsen layer equals the target surface temperature, $T_s$, contrary to the temperature of the heavy particles, $T_h$, which, according to the theory of the Knudsen layer, is 0.67 $T_s$. The estimation shows that there is no time for energy equilibration between the carbon atoms and the electrons in the Knudsen layer. Hence, it is assumed that the vapor is in the Saha equilibrium at a temperature, $T_e = T_s$, and $\frac{T_e}{T_h} = 1.5$.

Since the energy of the laser beam is supplied to electrons, the electron temperature will always exceed the temperature of heavy particles during the laser pulse. Therefore, the temperature ratio $\frac{T_e}{T_h} = 1.5$ is kept for first 9 ns of calculations. The temperature then reaches 25 K and the electron density Ne $\approx 1 \times 1026 \, m^{-3}$, and the ratio $\frac{T_e}{T_h}$ tends to unity.

After the cessation of the laser pulse, the energy equilibration time between electrons and heavy particles is a few nanoseconds. The energy source term IL was used in the form which fits the shape of the laser pulse and the plasma absorption coefficient included all possible absorption mechanisms, which are the electronatom inverse bremsstrahlung, the electron-ion inverse bremsstrahlung, the photoionization, and the Mie absorption.

The behavior of carbon plasma has been studied from the second half of the nineteenth century, in the form of astronomical observations of comet spectra. Modern research on carbon plasmas may help in understanding fullerene and carbon nanotube production processes, their kinetics, and molecular mechanisms.

The subsequent forms of condensed phase carbon nanostructures are formed. As theory and experiment move forward in a synergistic and complementary way, electronic and vibrational properties probed with modern computer simulation methods promise insights into carbon dynamics.

To some extent, the LIBS technique owes its widespread use to these carbon plasma applications. The LIBS method may be used to generate plasmas from any sample forms and phases. The emission spectra of these plasmas can be recorded with fast time resolution, and these spectra are then used to study the time evolution of plasma composition and temperature at nano- or picosecond resolution levels.

The LIBS method has been named this way due to the fact that laser photons ionize the media, and in these ionized media, volume electric discharges occur via dielectric breakdown processes. The components of the plasma are usually not in thermodynamic equilibrium and are usually highly excited.

## 3.3  Experimental Setup

Figure 3.6 shows the schematic diagram of the experimental setup used for generating carbon plasma. The experiment was conducted in a stainless steel vacuum chamber with different ambient environments and pressures. The setup consisted of a Q-switched Nd:YAG pulsed laser with a wavelength of 1064 nm, an energy of 740 mJ, a pulse width of 6 ns, and a frequency of 10 Hz. To focus the laser light through a quartz window onto a 4N pure graphite target, a plano-convex lens with a positive focus length of 18 cm was utilized.

Initially, the graphite target was irradiated with 30 laser shots. The resulting carbon plasma was then detected using a CR-39 detector, which was positioned perpendicular to the laser beam and adjusted with respect to the target using an electronically controlled translation stage. The detector was placed parallel to the substrate at a distance of 5 cm from the target.

The properties of the generated carbon plasma can be significantly influenced by the choice of ambient environment and pressure. The plasma composition, electron temperature, and density can all vary depending on these factors. In this experiment, the ambient environments tested were air and hydrogen, while the pressures ranged from 0.1 to 100 mbar.

When carbon plasma is generated in air, it typically contains a higher concentration of oxygen and nitrogen compared to plasma generated in a pure hydrogen environment. This is because oxygen and nitrogen are present in air, and they can be ionized during plasma generation. However, plasma generated in hydrogen has been

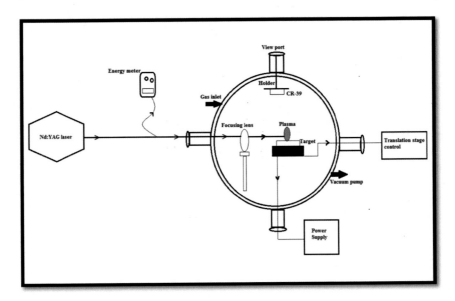

**Fig. 3.6** Schematic diagram of the experimental setup

shown to have a higher electron temperature and density than plasma generated in air. This is because hydrogen has a lower ionization potential than air, which allows for more efficient ionization and heating of the plasma.

The electron temperature of the plasma is an important parameter, as it determines the energy distribution of the electrons and can affect the ionization rate and the production of various plasma species. The electron density is also important as it affects the transport and recombination of charged particles in the plasma. Therefore, by varying the ambient environment and pressure, we can control the plasma parameters and tailor the plasma properties to suit specific applications.

In addition to the choice of ambient environment and pressure, the laser parameters also have a significant impact on the properties of the generated carbon plasma. For instance, varying the laser energy or pulse width can affect the plasma composition, electron temperature, and density. Higher laser energies can lead to more efficient ionization and heating of the plasma, resulting in a higher electron temperature and density. Similarly, shorter pulse widths can lead to more efficient ionization and heating of the plasma, resulting in a higher electron temperature and density.

The use of a 4N pure graphite target in this experiment is significant as it allows for the generation of carbon plasma with a high degree of purity. This is important for various applications, such as the production of carbon nanotubes and graphene, where impurities can have a significant impact on the properties of the final product. In addition, the use of a pure graphite target ensures that the carbon plasma generated in this experiment is of high quality and suitable for further analysis.

It is worth noting that the purity of the target material and the laser parameters used are important considerations for generating high-quality plasma. The experimental setup and conditions must be optimized to generate plasma with the desired properties, which can be further analyzed and utilized for various applications in materials science and plasma physics.

The experimental setup depicted in Fig. 3.6 offers a valuable platform for investigating the characteristics of carbon plasma in diverse ambient environments and pressures. Through the manipulation of these factors, along with the laser parameters, a comprehensive comprehension of the fundamental processes underlying plasma generation and the associated properties can be attained. This knowledge can then be utilized to optimize plasma-based procedures aimed at generating carbon-related materials with precise properties and functions.

The use of CR-39 detectors in this experiment allows for the measurement of the energy of carbon ions with high precision and accuracy. The etching process using NaOH solution is a widely accepted method for removing material from the ion tracks and preparing the detectors for analysis. By analyzing the diameter of the ion tracks using an optical microscope and applying $E = (D)^{6.16}/1.0864$, it is possible to estimate the energy of the carbon ions with good accuracy. Here, $D$ is the track diameter of ions in $\mu$m, and $E$ is the energy of ions in keV.

The use of multiple samples exposed to carbon ions under different conditions allows for a comprehensive study of the effect of ambient environment and pressure on the energy of the ions. By analyzing the data obtained from the samples, it is possible to identify trends and correlations between the ion energy and the experimental parameters. This information can be useful in optimizing plasma-based processes for the production of carbon-related materials with specific properties and applications.

# Chapter 4
# Production of Carbon Nanotube Using Arc Discharge Plasma

## 4.1 Introduction

Laser-induced plasma has garnered considerable attention due to its unique properties and potential applications. The use of Nd:YAG lasers in generating carbon plasma has been shown to be effective and offers precise control over plasma parameters. However, the dynamics of carbon ions in different ambient environments and pressures remain unclear.

In this book, the author sought to investigate the behavior of carbon ions in different ambient environments and pressures using CR-39 detectors. A Q-switched Nd:YAG pulsed laser was used to generate laser-induced carbon plasma in a stainless steel vacuum chamber. Plano-convex lenses were utilized to focus the laser beam on a pure graphite target, and the resulting plasma was detected using CR-39 detectors.

The CR-39 detectors were exposed to carbon ions under different ambient environment and pressures. After exposure, the detectors were etched using NaOH solution to reveal the ion tracks. The energy of the carbon ions was estimated by measuring the diameter of the ion tracks using an optical microscope. This allowed for a more precise understanding of the dynamics of carbon ions in different environments.

The results of this study provide significant insights into the behavior of carbon ions under varying environmental conditions. These findings have important implications for various fields, such as materials science and plasma physics, as they contribute to a better understanding of carbon ion dynamics. The ability to control carbon ions' behavior in various environments is critical in advancing technological applications in these fields.

Overall, this study's experimental design, which employed a Q-switched Nd:YAG pulsed laser, CR-39 detectors, and NaOH solution etching, allowed for a more comprehensive analysis of carbon ion behavior. Future research can build upon these findings to further advance our understanding of laser-induced plasma and its applications in various fields.

© The Author(s), under exclusive license to Springer Nature Singapore Pte Ltd. 2023
S. Daud, *Carbon Nanotubes*, SpringerBriefs in Applied Sciences and Technology,
https://doi.org/10.1007/978-981-99-4962-5_4

## 4.2  Carbon Ions Produce by Arc Discharge Plasma

The experimental setup utilized arc discharge plasma to produce carbon ions. This was achieved by placing two parallel graphite electrodes in a vacuum chamber and applying a high voltage between the anode and cathode. The resulting spark generated carbon plasma, which was directed toward the CR-39 track detector.

The CR-39 samples were exposed to carbon ions in both air and hydrogen at different pressures (0.1, 1, 10, and 100 mbar). As shown in Fig. 4.1a, b, the exposed samples exhibited varying levels of etching, indicating differences in the energy of the carbon ions. These differences in etching patterns can be attributed to variations in the energy levels of the carbon ions and the environmental conditions under which they were produced.

The observed differences in etching patterns can provide insights into the properties of the carbon ions, including their energy levels and compositions. Such information can be useful for optimizing the production of carbon ions for various applications, such as in the fabrication of carbon-based nanomaterials.

Figure 4.2a, b provide a closer look on the etching patterns of the CR-39 samples exposed to carbon ions in air and hydrogen at 0.1 mbar pressure, respectively. The burned surface of the CR-39 material indicates the presence of high-energy carbon ions. These images demonstrate the sensitivity of the CR-39 detector to the energy levels of carbon ions.

In addition to estimating the energy of the carbon ions, the CR-39 detector can also provide information about the spatial distribution of the ions. This is because

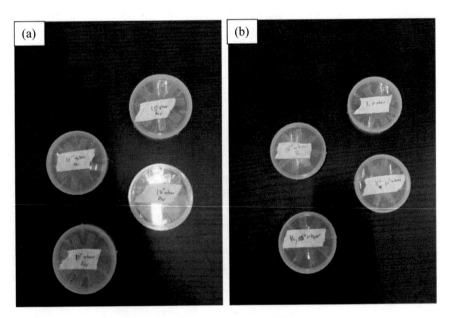

**Fig. 4.1** CR-39 samples in **a** air; **b** hydrogen at different pressures

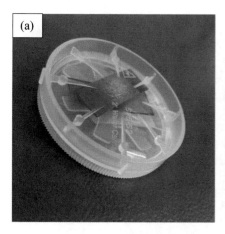
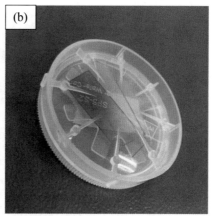

**Fig. 4.2** Enlarged CR-39 samples exposed in **a** air; **b** hydrogen at 0.1 mbar pressure

the diameter of the ion track is related to the angle of incidence of the ion, which in turn can provide information about the trajectory of the ion. By analyzing the spatial distribution of carbon ions, it is possible to gain insights into the plasma dynamics and the mechanisms behind ion production.

The use of arc discharge plasma for generating carbon ions has several advantages over other methods. For instance, it is a relatively simple and cost-effective method that can be easily scaled up for industrial applications. Additionally, it can produce a high flux of carbon ions, which is desirable for applications such as ion implantation and material processing. Overall, the results of this study highlight the potential of arc discharge plasma as a promising method for generating carbon ions with tunable energy levels and spatial distribution.

## 4.3 Etching Process of CR-39

The CR-39 detector is essential in analyzing the behavior of carbon ions generated by arc discharge plasma. To make the carbon ion tracks visible for further analysis, the detector undergoes an etching process. This process dissolves the non-exposed parts of the CR-39 surface, leaving only the ion tracks behind. In this experiment, 6.25 M NaOH solution has been used and maintained at a temperature of ($70 \pm 1$) °C for seven hours for the etching process. The setup for etching process is shown in Fig. 4.3.

The NaOH solution used in the etching process is highly corrosive and requires careful handling to prevent damage to the CR-39 detector. To ensure reliable results, it is crucial to follow the appropriate etching protocol and control the etching time

**Fig. 4.3** Etching process of CR-39 samples in NaOH solution

and temperature carefully. Based on literature, the optimal etching time for the CR-39 detector is between six and eight hours, depending on the concentration of the NaOH solution and the temperature used.

Once the etching process is complete, the CR-39 detector is removed from the NaOH solution and rinsed thoroughly with distilled water to remove any remaining solution. It is then left to dry before analysis under an optical microscope. The optical microscope allows for the observation and measurement of the diameter and density of carbon ion tracks on the detector surface. This information is critical in determining the energy and distribution of carbon ions produced by arc discharge plasma.

In this experiment, the CR-39 samples underwent optical microscope analysis to observe the carbon ion tracks after the etching process. The results as depicted in Fig. 4.4a, b show the CR-39 samples after exposed in air and hydrogen, respectively, after seven hours of etching in the NaOH solution. The clear and visible ion tracks evident in the images indicate a successful etching process, suggesting the production of carbon ions under the experimental conditions.

To summarize, the etching process plays a crucial role in analyzing the behavior of carbon ions produced by arc discharge plasma using CR-39 detectors. Proper adherence to the etching protocol is critical to obtaining reliable and high-quality results. Optical microscopes are used to analyze the diameter and density of ion tracks, providing valuable information about the energy and distribution of carbon ions.

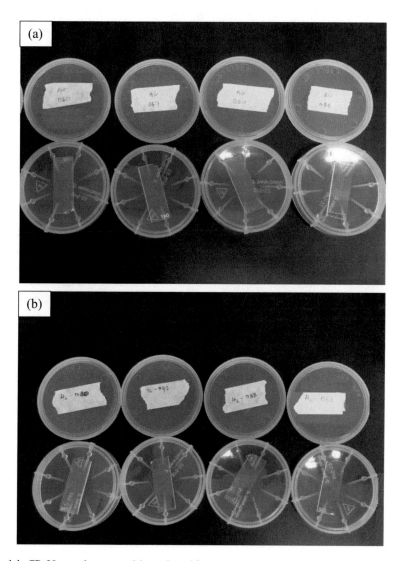

**Fig. 4.4** CR-39 samples exposed in **a** air and **b** hydrogen under different pressures after being etched in NaOH solution

## 4.4  Diameter of Carbon Ion

As previously mentioned, the experiment involved exposing CR-39 detectors to both hydrogen and air environments. After the etching process, ion tracks were observed and data was collected. An ion is an atom that has lost or gained electrons, resulting in a positive or negative charge. In contrast, neutral atoms or molecules have equal numbers of electrons and protons.

Carbon ions, which are carbon atoms that have lost one or more of their six electrons, are used in ion beam therapy, similar to protons. Heavy ions are charged particles that are heavier than helium. Carbon ions have been used in clinical settings for cancer treatment. Compared to protons, carbon ions have 12 times greater mass. The diameter of ion tracks was measured using an optical microscope, and ions with different diameters produced tracks with varying energies.

The CR-39 samples were exposed to ionizing radiation in air, resulting in the formation of ion tracks on their surface. These ion tracks were observed under a microscope and imaged for further analysis. To measure the diameter of the ion tracks, the AmScope optical microscope's diameter measurement tool was used.

Figure 4.5a–d depict the resulting images of ion tracks on the surface of CR-39 samples exposed to ionizing radiation in air, with each image corresponding to a different pressure level. Analyzing these images and measuring the ion track diameter can provide valuable information about the properties and nature of the ionizing radiation.

In this experiment, images of ion tracks on the surface of CR-39 samples were captured under an AmScope optical microscope at different pressures, and two examples were depicted in Fig. 4.6a, b, respectively. The results showed that the CR-39 material exhibited numerous tiny ion tracks, with diameters ranging from a few nanometers to several tens of nanometers, depending on the pressure at which the measurements were taken.

The precise measurement of these ion tracks is critical for a range of applications, including radiation dosimetry, particle physics, and materials science. For instance, in radiation dosimetry, the measurement of ion tracks can help to determine the amount of ionizing radiation received by the material, and this information is crucial for ensuring safety in various fields, including medical, nuclear, and space exploration. In particle physics, ion tracks can provide insights into the properties and behavior of subatomic particles, and in materials science, the study of ion tracks can help understand the effects of ionizing radiation on materials and their properties.

Table 4.1 displays the calculated average diameters of ion tracks in air at different pressures. The results demonstrate that the ion tracks' diameter is influenced by the pressure of the surrounding gas. The largest diameter of 18.29 μm was recorded at the lowest pressure of 0.1 mbar, whereas the smallest diameter of 7.72 μm was observed at the highest pressure of 100 mbar. These findings indicate that lower pressures lead to larger ion track diameters, while higher pressures result in smaller diameters in the CR-39 material.

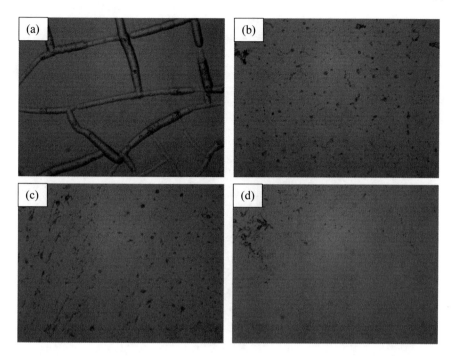

**Fig. 4.5**  Image of ion track exposed in air at **a** 0.1; **b** 1; **c** 10; and **d** 100 mbar pressures

The changes in ion track diameter observed at varying pressures can be attributed to differences in energy deposition and track formation mechanisms of ionizing particles. Higher pressures lead to increased ion-gas molecule collisions, resulting in energy loss and reduced ionization density in the material. This effect ultimately leads to smaller, less well-defined ion tracks. Conversely, lower pressures provide ions with a longer mean free path, allowing them to travel further before losing their energy. This results in higher ionization densities and larger, more defined ion tracks.

The measurement of ion track diameter provides valuable information about the properties of the materials and ionizing particles under study, making it a crucial tool for numerous scientific and technological applications.

The same method and software utilized to capture and measure the diameter of ion tracks in air were applied to analyze CR-39 carbon ions in hydrogen. The diameter and images of ion tracks were captured and measured at different pressures using the same approach.

Figure 4.7a–d depict images of ion tracks observed in a hydrogen environment at pressures of 0.1, 1, 10, and 100 mbar, respectively. These images demonstrate that the ion tracks created by the passage of high-energy particles in hydrogen have a similar appearance to those observed in air, consisting of small, well-defined tracks.

The diameter of ion tracks in hydrogen is influenced by the pressure of the surrounding gas, where higher pressures result in smaller diameter tracks due to reduced ionization density. Conversely, lower pressures lead to higher ionization

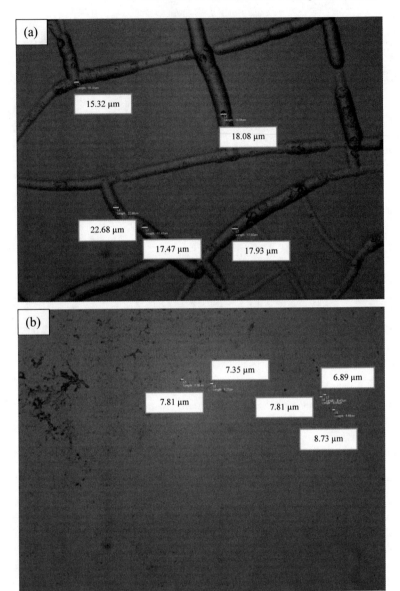

**Fig. 4.6** Image of CR-39 ion track at **a** 0.1 and **b** 100 mbar pressure in air

densities and larger ion tracks. Precisely measuring the diameter of ion tracks in hydrogen, similar to air, is crucial for numerous applications, such as radiation dosimetry, particle physics, and materials science.

Notably, studying ion tracks in hydrogen has unique advantages over air. Hydrogen has a lower atomic number and higher cross-section for ionization, making it easier

**Table 4.1**  Diameter of the ion tracks in air

| Air | | | | |
|---|---|---|---|---|
| No. of sample | Pressure, $p$ (mbar) | | | |
| | 0.1 | 1 | 10 | 100 |
| Sample 1 | 18.08 | 16.09 | 16.55 | 7.81 |
| Sample 2 | 22.68 | 18.39 | 16.55 | 7.35 |
| Sample 3 | 17.93 | 20.68 | 13.48 | 6.89 |
| Sample 4 | 17.47 | 17.01 | 13.48 | 7.81 |
| Sample 5 | 15.32 | 18.39 | 12.26 | 8.73 |
| Average ($\mu$m) | 18.29 | 18.11 | 14.46 | 7.72 |

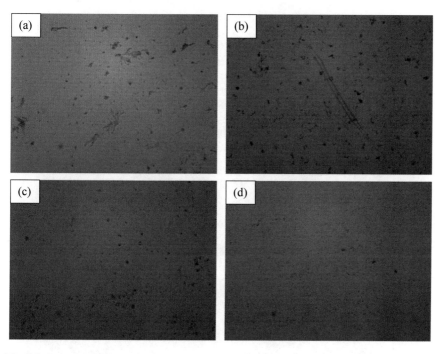

**Fig. 4.7**  Image of CR-39 ion track exposed in hydrogen at **a** 0.1; **b** 1; **c** 10; and **d** 100 mbar pressure

to detect and analyze the ion tracks created by high-energy particles. Furthermore, studying ion tracks in hydrogen can provide valuable information about the properties of the particles, including their energy, charge, and trajectory.

Figure 4.8a, b present the measured diameters of ion tracks in hydrogen at different pressures, showcasing the diameters of ion tracks observed at 0.1 and 100 mbar, respectively. As expected, the diameter of ion tracks in hydrogen exhibits a similar trend to those observed in air, where lower pressures result in larger diameters and higher pressures lead to smaller diameters.

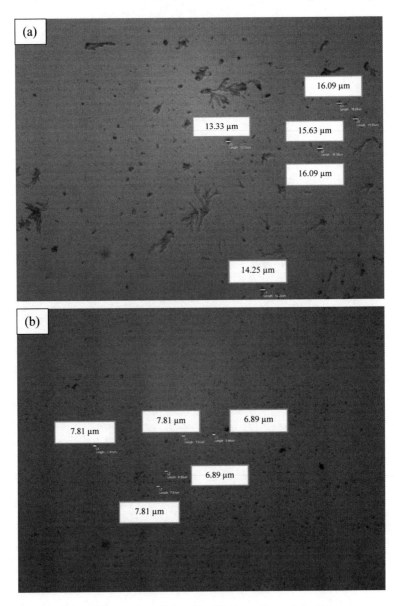

**Fig. 4.8** Image of CR-39 ion track at **a** 0.1 and **b** 100 mbar pressure in hydrogen

Interestingly, the average diameter of ion tracks measured in hydrogen at 0.1 mbar was larger than that observed at 100 mbar, which aligns with the trend observed in air. Furthermore, the study of ion tracks in hydrogen offers unique advantages due to its lower atomic number and higher cross-section for ionization, making it an ideal medium for the analysis of high-energy particles.

   Measuring the diameter of ion tracks in hydrogen, as in air, can provide valuable information about the properties of the ionizing particles and the materials being studied, making it a crucial tool for numerous scientific and technological applications, such as radiation dosimetry, particle physics, and materials science.

   Table 4.2 presents the average diameter of ion tracks in hydrogen at different pressures, which was calculated to ensure accurate measurements. The results demonstrate a similar trend to that observed in air, where the diameter of ion tracks decreases as the pressure of the surrounding gas increases. Notably, the largest diameter of ion tracks observed in hydrogen was 15.08 μm at the pressure of 0.1 mbar, while the smallest diameter was 7.44 μm at the pressure of 100 mbar. These findings align with the trend observed in air, where the smallest diameter of ion tracks was also observed under the highest pressure.

   The precise measurement of the diameter of ion tracks in hydrogen, like in air, plays a critical role in several scientific and technological applications, such as radiation dosimetry, particle physics, and materials science. These results provide significant insights into the behavior of ionizing particles in different environments, thereby enhancing our understanding of various processes that occur in nature.

   Figure 4.9 features a bar chart that presents a comparison of the diameter of ion tracks against pressure in both air and hydrogen environments. The chart clearly shows that, irrespective of the environment, the diameter of ion tracks generally decreases as the pressure of the surrounding gas increases. This finding highlights the importance of considering the pressure of the surrounding gas in studies involving ionizing particles and their tracks.

   The comparison of the diameter of ion tracks between air and hydrogen revealed a notable difference, as can be seen from the bar chart presented in Fig. 4.9. Specifically, the average diameter of ion tracks in air was found to be larger than that in hydrogen across all pressures. The diameter values of CR-39 ion track at 0.1, 1, 10, and 100 mbar were calculated to be 18.29 μm and 15.08 μm, 18.11 μm and 14.25 μm, 14.46 μm and 13.51 μm, and 7.72 μm and 7.44 μm, respectively.

   These results shows that the behavior of ionizing particles differs between air and hydrogen environments, resulting in the observed differences in the diameter of ion

**Table 4.2**  Diameter of the ion track in hydrogen

| Hydrogen | | | | |
| --- | --- | --- | --- | --- |
| No. of sample | Pressure, $p$ (mbar) | | | |
| | 0.1 | 1 | 10 | 100 |
| Sample 1 | 16.09 | 13.79 | 12.87 | 7.81 |
| Sample 2 | 15.63 | 14.25 | 13.79 | 7.81 |
| Sample 3 | 16.09 | 15.63 | 13.79 | 6.89 |
| Sample 4 | 13.33 | 15.17 | 13.79 | 6.89 |
| Sample 5 | 14.25 | 12.41 | 13.33 | 7.81 |
| Average | 15.08 | 14.25 | 13.51 | 7.44 |

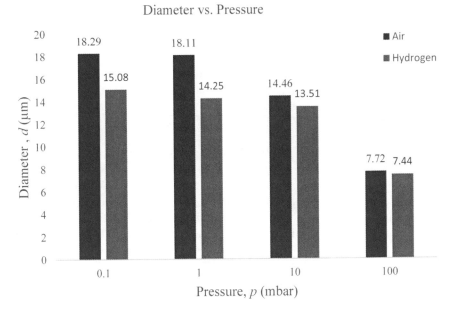

**Fig. 4.9** Graph ion tracks diameter against pressure in air and hydrogen

tracks. The measurement and analysis of ion tracks offer valuable insights into the interaction of ionizing particles with materials, and the discrepancies between air and hydrogen may have important implications for various scientific and technological applications.

## 4.5  Energy of Carbon Ion in Air and Hydrogen

The study of the resulting energy of carbon ions under different environmental conditions and pressures is important for understanding the behavior of high-energy particles and their interactions with materials. The production of carbon ions by arc discharge plasma is a widely used method for generating high-energy ions, and the surrounding gas can significantly affect the energy of the produced ions.

The mechanism behind the influence of environmental conditions on the formation of carbon ions is still not fully understood and may involve several factors, such as the ionization potentials and collision processes with gas molecules. However, the observed differences in the energy of carbon ions between air and hydrogen suggest that the ionization process is affected by the properties of the surrounding gas.

The finding that the energy of carbon ions in air is generally higher than that in hydrogen has important implications for various scientific and technological applications, such as ion implantation and ion beam lithography. The energy and characteristics of the ions used in these processes are crucial for achieving the desired performance of the materials being processed. Understanding the environmental factors that influence the energy of carbon ions can therefore help optimize these applications and improve their efficiency.

Figure 4.10 displays a graph that depicts the relationship between the energy of carbon ions and the gas pressure of the surrounding environment, in both air and hydrogen. The data reveals that across all pressures measured (0.1, 1, 10, and 100 mbar), the energy of carbon ions was consistently higher in air than in hydrogen. Specifically, the energy of carbon ions in air is measured to be 54.86 keV, 51.62 keV, 12.90 keV, and 0.27 keV at 0.1, 1, 10, and 100 mbar, respectively. As for the energy of carbon ion hydrogen, it was measured to be 6.71 keV, 11.78 keV, 1.08 keV, and 0.22 keV at 0.1, 1, 10, and 100 mbar, respectively.

This observation supports the notion that the properties of the surrounding gas play a crucial role in the ionization and formation of carbon ions and may be due to various factors such as ionization potentials and collision processes with gas molecules. These findings have significant implications for the use of carbon ions in a range of scientific and technological applications, where the energy and properties of the ions can significantly impact the performance of the materials being processed.

The disparity in the energy of carbon ions between air and hydrogen can be attributed to the composition and ionization potential of the respective gases. The presence of more complex oxygen and nitrogen molecules in air can potentially aid the ionization process, resulting in the production of higher energy ions. On the other hand, the simpler composition and lower ionization potential of hydrogen may constrain the ionization and energy of the carbon ions formed.

These findings underscore the significance of environmental conditions in the generation and manipulation of carbon ions, with implications for a diverse range of applications in fields such as materials science, chemistry, and biomedicine.

## 4.6  Energy of Carbon Ion in Different Pressure

To investigate the impact of pressure on the energy of carbon ions, the experiment was conducted in controlled environments of air and hydrogen, with precise pressure readings taken at four different levels (0.1, 1, 10, and 100 mbar). The energy of carbon ions was then measured for each pressure in both environments. The results clearly demonstrated that pressure of the gas had a significant effect on the energy of carbon ions produced, regardless of the environment. The pressure was systematically varied to determine the optimal pressure range for accurately measuring the energy of carbon ions.

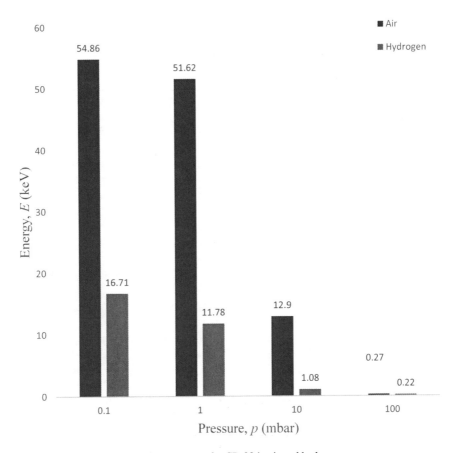

**Fig. 4.10** Graph of energy against pressure for CR-39 in air and hydrogen

## 4.6.1  Different Pressure in Air

The experiment was conducted in a controlled environment with four different air pressures. The energy of carbon ions produced by an arc discharge plasma was measured at each pressure, and the results were given in Table 4.3.

The data revealed that the energy of carbon ions decreases as the air pressure increases. The highest energy of carbon ions produced was recorded at the pressure of 0.1 mbar, with the energy value of 54.86 keV, while the lowest energy recorded was at the pressure of 100 mbar, with the energy value of 0.27 keV. The energy of carbon ion at 1 and 10 mbar pressure was recorded to be 51.62 keV and 12.90 keV, respectively.

**Table 4.3** Energy of carbon ions due to different pressures in air

| Air | |
| --- | --- |
| Pressure (mbar) | Energy (keV) |
| 0.1 | 54.86 |
| 1 | 51.62 |
| 10 | 12.90 |
| 100 | 0.27 |

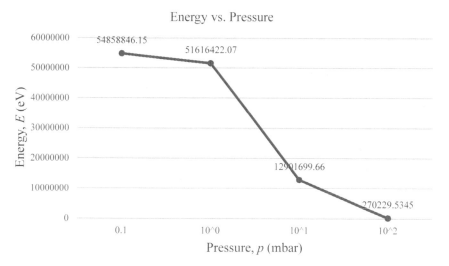

**Fig. 4.11** Graph of carbon ions energy against pressure in air

To visualize the relationship between pressure and the energy of carbon ions, a graph of energy versus pressure was plotted and shown in Fig. 4.11. The graph showed a decreasing trend in the energy of carbon ions as the air pressure increases. This trend can be attributed to the fact that at higher pressures, there are more gas molecules present that can interact with the carbon ions, causing energy loss through collisional processes.

### 4.6.2  Different Pressure in Hydrogen

The energy of carbon ions produced by an arc discharge plasma was then measured in a controlled environment with four different hydrogen pressures, and the results were given in Table 4.4.

The data showed that the energy of carbon ions decreases as the hydrogen pressure increases. The highest energy of carbon ions was recorded at a pressure of 0.1 mbar, with the energy value of 16.71 keV, while the lowest energy was recorded at the

**Table 4.4** Energy of carbon ions due to different pressures in hydrogen

| Hydrogen | |
| --- | --- |
| Pressure (mbar) | Energy (keV) |
| 0.1 | 16.71 |
| 1 | 11.78 |
| 10 | 1.08 |
| 100 | 0.22 |

pressure of 100 mbar, with the energy value of 0.22 keV. The energy of carbon ion at 1 and 10 mbar pressure was recorded to be 11.78 keV and 0.022 keV, respectively.

To visualize the relationship between pressure and energy of carbon ions in hydrogen gas, a graph of energy versus pressure was plotted and shown in Fig. 4.12. The graph displayed a decreasing trend in the energy of carbon ions as the hydrogen pressure increased. This trend can be attributed to the fact that at higher pressures, there are more hydrogen molecules present, which can interact with the carbon ions and cause energy loss through collisional processes.

The comparison of energy output between hydrogen and air environments, as well as the dependence of carbon ion energy on hydrogen pressure, provides valuable insights into the behavior of carbon ions in different environments. Such information can be utilized in a wide range of applications, including the development of plasma-assisted processes for materials synthesis and modification.

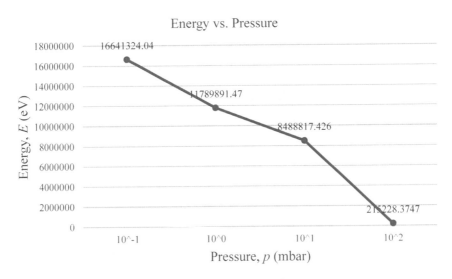

**Fig. 4.12** Graph of carbon ion energy against pressure in hydrogen

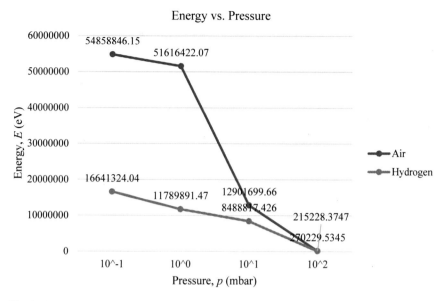

**Fig. 4.13** Graph of carbon ion energy against pressure in air and hydrogen

### 4.6.3 Comparison of the Energy of Carbon Ions Produced in Air and Hydrogen at Different Pressures

The energy of carbon ion under different pressures in air and hydrogen was determined and presented in Tables 4.3 and 4.4, respectively. The data was then used to plot a graph of energy against pressure for both environments, as shown in Figs. 4.11 and 4.12, respectively. These graphs were plotted together in the same chart as shown in Fig. 4.13 to allow a direct comparison between the energy of carbon ion in air and hydrogen that were influenced by pressure.

Based on the comparison made, it can be seen that the energy of carbon ion in air was higher than that in hydrogen at all pressures, and that the highest energy was obtained at the lowest pressure for both environments. This can be explained by the fact that at lower pressures, the distance between the electrodes is larger, which increases the electric field strength and thus the energy of the carbon ion.

From the graphs plotted, it can be observed that there is an inverse relationship between pressure and energy of carbon ion in both air and hydrogen environments. Specifically, the highest energy was observed at the lowest pressure in both environments. The inverse relationship between pressure and energy of carbon ion can be explained by the fact that as the pressure increases, there are more collisions between the carbon ions and the gas molecules in the environment, which leads to more energy losses due to ionization and excitation processes. As a result, the energy of the carbon ions decreases as the pressure increases.

**Table 4.5** Energy of carbon ion due to different pressure and environment

| Pressure (mbar) | Air | | Hydrogen | |
| --- | --- | --- | --- | --- |
| | Diameter ($\mu$m) | Energy (keV) | Diameter ($\mu$m) | Energy (keV) |
| 0.1 | 18.29 | 54.86 | 15.07 | 16.64 |
| 1 | 18.11 | 51.62 | 14.25 | 11.79 |
| 10 | 14.46 | 12.90 | 13.51 | 8.49 |
| 100 | 7.16 | 0.27 | 7.44 | 0.22 |

On the other hand, at lower pressures, there are fewer collisions between the carbon ions and the gas molecules, which allow the ions to retain more of their initial energy. Therefore, the highest energy was found at the lowest pressure in both air and hydrogen environments.

Table 4.5 presents the data for the energy of carbon ion at different pressures in air and hydrogen, which is plotted in a graph as shown in Fig. 4.13. The graph displays the decreasing trend of energy with increasing pressure in both environments.

At pressure 0.1 mbar, the energy of carbon ion in air was 54.86 keV, while at 100 mbar, it decreased to 0.27 keV. Similarly, in hydrogen, the energy decreased from 16.64 keV at 0.1 mbar to 0.22 keV at 100 mbar. The graph clearly shows that the energy of carbon ion decreases gradually as the pressure increases in both air and hydrogen environments.

# Chapter 5
# Conclusion

## 5.1 Conclusion

This chapter provides the summary and recommendations for extending work in the future. In conclusion, this book has delved into the fascinating world of carbon nanotubes, uncovering their remarkable properties, potential applications, and the cutting-edge research surrounding them. Through meticulous exploration and analysis, we have witnessed the unique structural characteristics of carbon nanotubes, their exceptional mechanical strength, electrical conductivity, and thermal stability. These properties have propelled them into the forefront of scientific and technological advancements, holding great promise for revolutionizing various fields.

Throughout the chapters, we have explored the diverse applications of carbon nanotubes, ranging from electronics and energy storage, to biomedical engineering and materials science. Their exceptional conductivity has enabled advancements in nanoelectronics and has the potential to drive the development of faster, smaller, and more efficient devices. Their high aspect ratio and strength have paved the way for breakthroughs in materials engineering, leading to the creation of lightweight, yet strong composites with unparalleled mechanical properties.

Moreover, we have examined the synthesis and characterization techniques that enable the precise control and manipulation of carbon nanotubes at the nanoscale. From chemical vapor deposition to arc discharge and laser ablation, each method offers unique advantages and challenges, contributing to the ever-expanding knowledge base in this field.

## 5.2 Recommendation

As we conclude this journey, it is important to highlight the ongoing research and future prospects for carbon nanotubes. Scientists and engineers continue to push the boundaries of knowledge, exploring novel synthesis approaches, functionalization

S. Daud, *Carbon Nanotubes*, SpringerBriefs in Applied Sciences and Technology, https://doi.org/10.1007/978-981-99-4962-5_5

techniques, and integration strategies. The challenges of scalability, purification, and cost-effectiveness persist, but the potential impact of carbon nanotubes on various industries and technologies drives the relentless pursuit of solutions.

Ultimately, this book serves as a testament to the remarkable versatility and potential of carbon nanotubes. It is a call to researchers, entrepreneurs, and policymakers to embrace the possibilities that these nanoscale wonders offer and to work collaboratively toward realizing their full potential. As we stand on the precipice of a new era, driven by advancements in nanotechnology, the future of carbon nanotubes holds the promise of transforming our lives, revolutionizing industries, and shaping a more sustainable and technologically advanced.

# Index

Printed in the United States
by Baker & Taylor Publisher Services